Fundamentals of Coalbed Methane

Fundamentals of Coalbed Methane

Editor

Ashish Sharma

scitus
academics

Fundamentals of Coalbed Methane
Edited by **Ashish Sharma**

Printed in 2017

ISBN: 978-1-68117-376-4

Library of Congress Control Number: 2015941564

© 2016 by
SCITUS Academics LLC,
616, Corporate Way, Suite 2, 4766,
Valley Cottage, NY 10989

www.scitusacademics.com

Contents

Preface

We will discuss various aspect of technology for the evaluation and development of coalbed methane (CBM) reservoirs. This article discusses the gas storage and flow mechanism in CBM reservoirs, their differences with conventional gas reservoirs, and their impact on production behavior. In addition, the impact of mechanical properties of coal on CBM reservoirs is discussed. Coalbed methane (CBM) or coal-bed methane is a form of natural gas extracted from coal beds.In recent decades it has become an important source of energy in United States, Canada, Australia, and other countries. The term refers to methane adsorbed into the solid matrix of the coal. It is called 'sweet gas' because of its lack of hydrogen sulfide. The presence of this gas is well known from its occurrence in underground coal mining, where it presents a serious safety risk. Coalbed methane is distinct from a typical sandstone or other conventional gas reservoir, as the methane is stored within the coal by a process called adsorption. The methane is in a near-liquid state, lining the inside of pores within the coal (called the matrix). The open fractures in the coal (called the cleats) can also contain free gas or can be saturated with water.

Editor

Chapter 1

Effects of Igneous Intrusion on Microporosity and Gas Adsorption Capacity of Coals in the Haizi Mine, China

Jingyu Jiang[1, 2, 3] and Yuanping Cheng[1, 2]

[1]Faculty of Safety Engineering, China University of Mining & Technology, Xuzhou, Jiangsu 221116, China

[2]National Engineering Research Center for Coal & Gas Control, China University of Mining & Technology, Xuzhou 221008, China

[3]Post-Doctoral Research Station, Yima Mining Corporation Limited, Sanmenxia, Henan 472300, China

ABSTRACT

This paper describes the effects of igneous intrusions on pore structure and adsorption capacity of the Permian coals in the

Huaibei Coalfield, China. Twelve coal samples were obtained at different distances from a ~120 m extremely thick sill. Comparisons were made between unaltered and heat-affected coals using geochemical data, pore-fracture characteristics, and adsorption properties. Thermal alteration occurs down to ~1.3 × sill thickness. Approaching the sill, the vitrinite reflectance (R_o) increased from 2.30% to 2.78%, forming devolatilization vacuoles and a fine mosaic texture. Volatile matter (VM) decreased from 17.6% to 10.0% and the moisture decreased from 3.0% to 1.6%. With decreasing distance to the sill, the micropore volumes initially increased from 0.0054 cm^3/g to a maximum of 0.0146 cm^3/g and then decreased to 0.0079 cm^3/g. The results show that the thermal evolution of the sill obviously changed the coal geochemistry and increased the micropore volume and adsorption capacity of heat-affected coal (60–160 m from the sill) compared with the unaltered coals. The trap effect of the sill prevented the high-pressure gas from being released, forming gas pocket. Mining activities near the sill created a low pressure zone leading to the rapid accumulation of methane and gas outbursts in the Haizi Mine.

INTRODUCTION

Many coalbed methane (CBM) basins such as the Raton and San Juan basins in the USA, the Gunnedah Basin in Australia, and the Qinshui and Fuxin basins in China have undergone contact metamorphism or thermal maturation directly or indirectly related to igneous intrusions [1–3]. The igneous intrusions provide a high-temperature and high-pressure environment for coal seams, which promotes the thermal evolution of the coal seam and speeds up the generation of gas [4]. The effects of localized igneous intrusions on the coal rank, petrology, geochemical, microstructure, and adsorption-desorption characteristics have been studied in numerous papers [5–10]. Two igneous sills in the Gunnedah Basin in Australia had positive effects on the gas content of CBM [2]. More recently, [11] suggested that the heating effect of a dike had enhanced not only the adsorption and porosity of metamorphosed coals but also the gas diffusivity and trap capacities of gas storage. Contrary to the opinion of [6, 11] indicated that the intrusion dikes strongly decreased the coal mesopore, micropore volume, porosity, and surface area that may have negative effects on gas migration in coalbeds adjacent to the dike in the Illinois Basin. The influences of igneous intrusions on coal pores and fractures vary significantly depending on the intrusion patterns, the coal ranks after the intrusion, and the nature of the adjacent formation surrounding the intrusion [10]. However, studies on the influences of localized intrusions on coal pore structure and adsorption capacity are still insufficient.

Igneous intrusions thermally and geochemically alter coal, often causing safety problems for underground coal mines [12]. Coal and gas outburst are dynamic disasters which may result in the projection of fragmented coal rock and rapid release of gases from the working face [13]. Five methane outburst disasters associated with dikes and sills occurred in the coal mines in Highveld coalfield of South Africa in the early 1990s [14]. The enhanced methane storage capacity and lower diffusivity of dike material can partly explain the occurrence of gas pockets encountered in the vicinity of intrusions [11]. Since the early 1980s, fifteen methane outburst disasters were reported to be associated with sills in underground mines in the Huaibei Coalfield of China. Three affected underground mines were the Haizi, Shitai, and Wolonghu Mines [15]. Magmatic activity occurred frequently in the Haizi Mine which has suffered eleven gas outbursts and one water inrush accident under a thick-hard igneous rock [4]. However, the effects of sill intrusions on coal adsorption capacity and their implications for methane outbursts have not been fully understood. The purpose of this paper is to study the effects of localized igneous intrusions on coal microporosity and adsorption capacity and their implications for methane accumulation and gas outburst.

GEOLOGICAL BACKGROUND

The Huaibei Coalfield, located in the northern Anhui Province in China, has three mining areas: Suixiao, Suxian, and Linhuan. The

Huaibei Coalfield is one of the country's major coalfields, with 23 active underground coal mines. The coals of Permian age are used mainly for power generation [16]. The effect of igneous intrusions on coal geochemistry, adsorption capacity, and the CBM generation and accumulation in Huaibei Coalfield was macroscopically investigated [10, 17]. The Haizi Mine, belonging to the Linhuan mining areas, is located in the northwestern Huaibei Coalfield and covers 33.8 km². The number 7, 8, 9, and 10 coal seams are the main mining layers. The locations of the study area are shown in Figure 1.

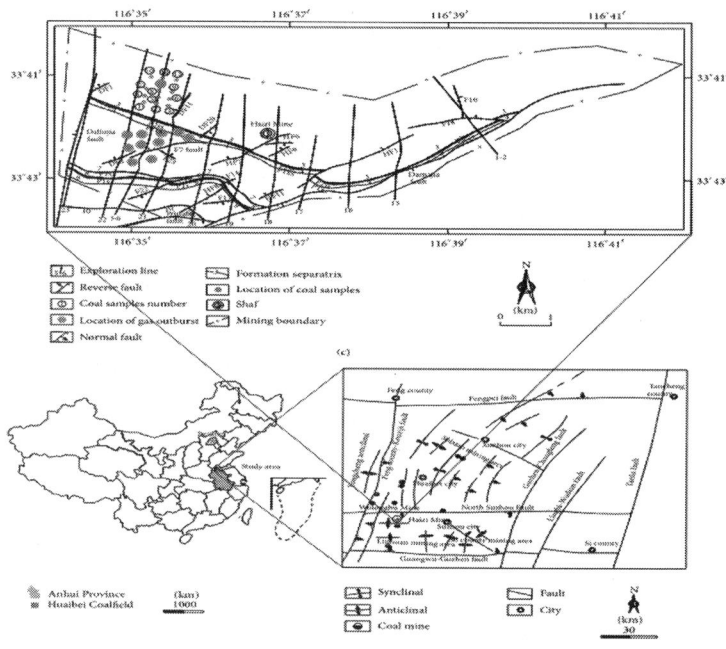

Figure 1: Map showing study area in China (a), geologic structure of the Huaibei Coalfield (b), and data locations of the coals from Haizi Mine (c).

The Cretaceous Yanshanian magmatic intrusions were widespread in the Haizi Mine, and the number 1 to 10 coal seams were invaded by igneous rock without exception. The diorite sill invades along the number 5 coal seam in the western Haizi Mine, 6.5 km long with a persistent thickness of 120 m [4]. The profile of the extremely thick sill (ETS) and coal seams disclosed by boreholes is shown in Figure 2. The northern part of the Haizi Mine adjoins the North Suzhou Fault which is one of main regional tectonics in the Huaibei Coalfield. The North Suzhou Fault, which is 240 km long and 4–6 km wide, formed in the late Paleozoic coal-forming period [18]. In the Yanshan Period magma upwelled through the North Suzhou Fault, invading the Haizi Mine. The Daliujia and the other faults controlled the direction and distribution of the igneous intrusions (Figure 1).

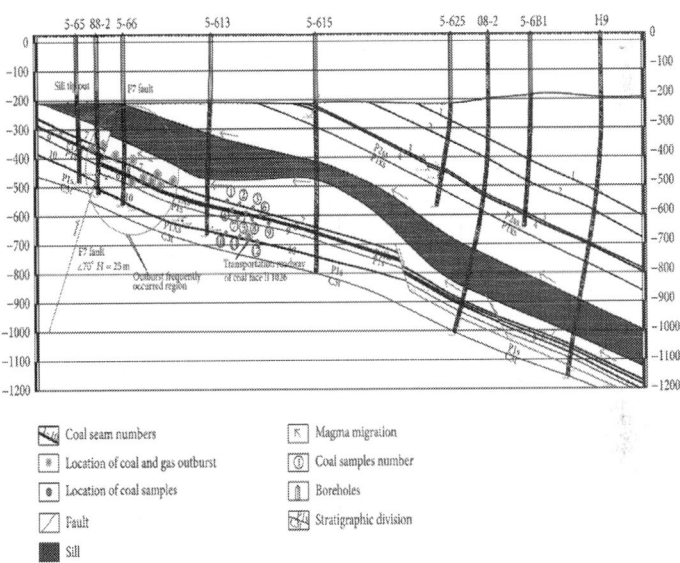

Figure 2: Profile of sill intrusions and coal seams disclosed by boreholes from Haizi Mine.

COAL SAMPLES AND ANALYTICAL PROCEDURES

Twelve fresh coal samples were sampled from coal seams number 7, 8, 9, and 10 under the ETS of the Haizi Mine. Three samples from each coal seam were taken at different distances (60–200 m) from the sill (Figure 2). The proximate analysis followed ASTM standards [19]. Random vitrinite reflectance (R_o) measurements were performed using a microscope photometer (Zeiss, Germany) according to international standards [20]. The maceral composition of the coals was determined by incident light microscopy and oil immersion according to international standards [21]. Scanning electron microscopy (SEM) analyses were conducted using a HITACHI S-3000N on selected samples. The BET specific surface was obtained using nitrogen gas as the adsorptive at the boiling point temperature of liquid nitrogen, whereas micropore volume, micropore surface area, and average micropore diameter were determined using low-pressure CO_2 at 273 K (AUTOSORB-1, Quantachrome Instruments Co., USA). We tested the methane adsorption isotherm of dry coal for determining the methane adsorption capacity in coal according to the standard of Chinese coal industry, MT/T752-1997. Crushed coal samples of ~50 g and 0.2–0.25 mm particle size were exposed to gas pressures of up to ~5 MPa at 30°C. The adsorption volume can be obtained using the methods detailed in previous research [22]. The initial gas-releasing rate index, ΔP, was obtained according to standards [23]. Gas adsorbed by crushed coal samples of 3.5 g of 0.2–0.25 mm particle size at a pressure of 0.1 MPa was released into a fixed vacuum space. The pressure rise in the space for the period of 10–60 s after the release, ΔP (in mmHg), is an index representing the methane release capacity of the coal. The coal firmness coefficient f was measured using a dropping hammer according to Chinese coal industry standard, MT 49–1987.

RESULTS AND DISCUSSION

Coal Rank Analyses

The R_o increases from levels of 2.30% (as measured by oil immersion) to 2.78% (Table 1).

Table 1: Geochemical and petrographic analyses of Haizi coal samples

| Sample | CN | D(m) | Proximate analysis (wt.%) | | | Macerals (vol.%) | | | R_o (%) | T(°C) |
			Mois	Ash	VM	V	I	M		
Number 1	7	60	1.6	45.6	10.0	83.4	14.2	2.4	2.78	283
Number 2	7	70	1.6	36.4	10.2	85.3	12.7	2.0	2.75	282
Number 3	7	80	1.7	38.8	10.5	85.7	12.0	2.3	2.71	280
Number 4	8	90	2.6	20.5	11.2	87.3	11.3	1.4	2.69	279
Number 5	8	100	2.4	20.0	11.6	87.6	10.5	1.9	2.60	275
Number 6	8	110	2.3	22.5	11.2	87.2	10.8	2.0	2.65	277
Number 7	9	120	2.5	18.7	11.8	86.4	11.2	2.4	2.59	274
Number 8	9	130	2.5	21.5	12.9	87.2	10.6	2.2	2.62	275
Number 9	9	140	2.6	20.1	12.2	85.6	12.0	2.4	2.53	271
Number 10	10	160	3.0	11.4	15.6	86.4	9.8	3.8	2.30	259
Number 11	10	180	2.7	10.6	16.4	87.9	8.8	3.3	2.32	260
Number 12	10	200	2.5	9.2	17.6	88.2	8.3	3.5	2.41	265

Note: CN: coal seam number;D: distance from sill boundary; Mois: moisture; Ash is on a dry basis; VM: volatile matter, on dry ash free (daf) basis; V: vitrinite; I: inertinite; M: mineral.

R_o: random vitrinite reflectance;T: estimated paleotemperature.

Approaching the sill, the R_o curve shows a slight decrease from coal sample 12 (R_o =2.41%, ~200 m from the sill) of the number 10 coal seam to sample 10; then, there is a rapid increase from sample 10 (R_o =2.30%, ~160 m from the sill) to sample 1 (R_o =2.78%, ~60 m from the sill) of the number 7 coal seam (Figure 3).

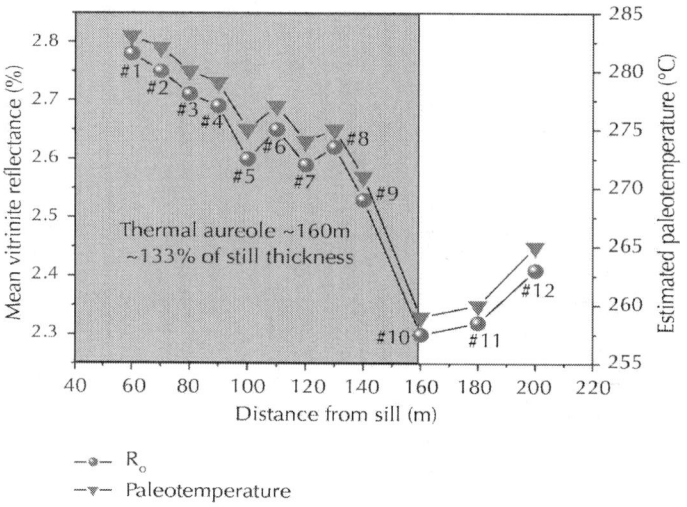

Figure 3: Variation in vitrinite reflectance and estimated paleotemperatures with distance from the sill.

These variations in R_o suggest that the thermal aureoles of the sill are approximately 160 m, ~1.33 × sill thickness. The slow growth in R_o from samples 10 to 12 follows the normal burial coalification tracks, the buried depth increases, R_o increases due to the geothermal metamorphism (Figure 3), and we therefore believe that samples 11 (R_o =2.32%, ~180 m from the sill) and 12 (R_o =2.41%, ~200 m from the sill) were not within the scope of the thermal aureoles. However, samples 1 to 10 did not follow the normal burial coalification tracks, the buried depth increases, and R_o =2.32 decreased due to the thermal evolution of the ETS. So we believe that coal samples 1 to 10 (inflection point of R_o =2.32 curve) were within the scope of the thermal aureoles of the sill (Figure 3).

The paleotemperatures attained by the coal can be estimated from Barker and Pawlewicz's relationship [24]. The estimated temperatures for the Haizi coal data are plotted against the distance from the contact in Figure3. The thermal evolution temperature (°C) of the Cretaceous Yanshanian magmatic metamorphism to the number 7 coal seam of the Haizi Mine is estimated to be at least 283°C.

The macroscopic appearance of the igneous rock by sill intrusion is shown in Figure 4(a). The diorite sill was mostly gray, with some light brown sections. The same megascopic appearance of the sill intrusions was found in the nearby Wolonghu Mine of the Huaibei Coalfield. This result verifies that the diorite rock (approximately 145 Ma), which runs along the North Suzhou Fault [18], was the source of magma both in the Haizi and Wolonghu Mines. Contact metamorphism dramatically increased the rank of the Wolonghu coal (increased from 2.74% to 5.03%) at 0–60 m from the sill in the horizontal direction, while the thermal evolution of the sill in the Haizi Mine slightly increased it in the coal 60–200 m from the sill (vertically) from 2.30% to 2.78%.

(a)

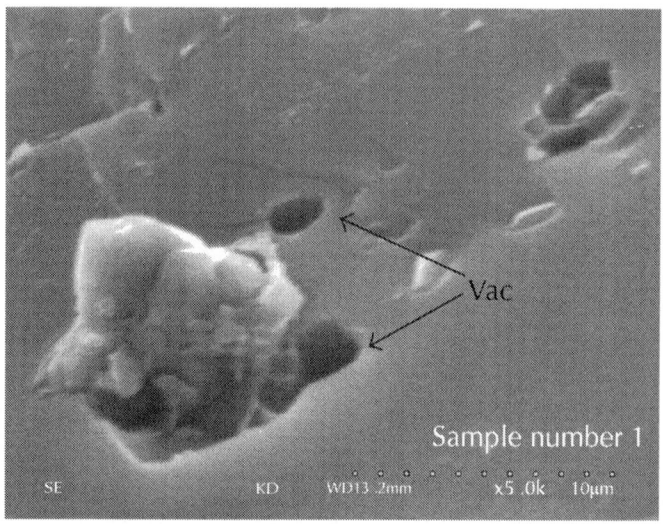

(b)

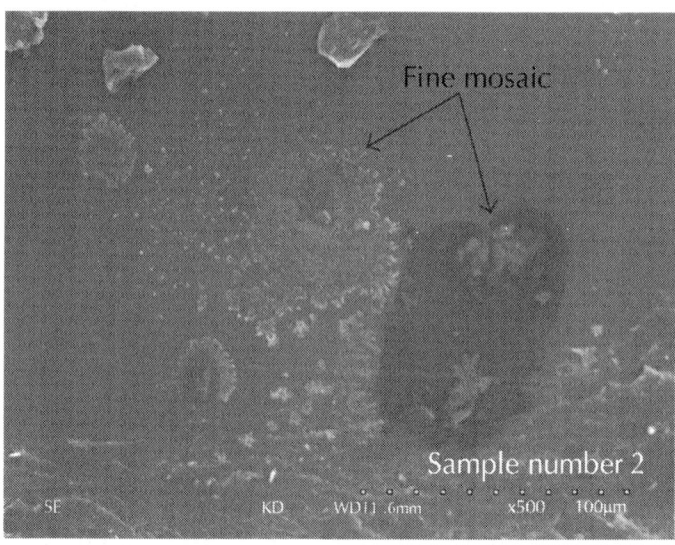

(c)

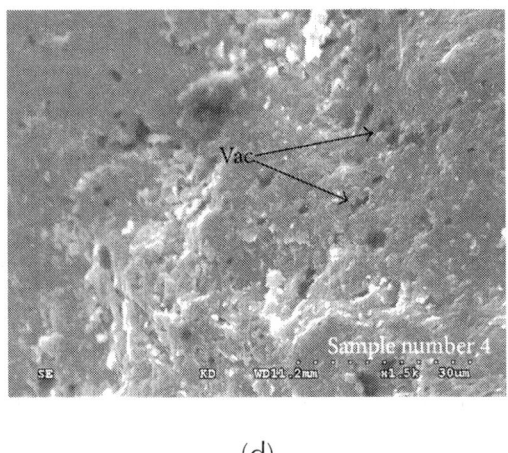

(d)

Figure 4: Macroscopic appearance of the igneous sills (a). Photomicrographs (b), (c), and (d) were obtained using SEM. Altered coal showing devolatilization vacuoles (vac), sample 1, R_o =2.78, magnification 5000x (b); altered coal showing fine mosaic texture, sample 2, R_o =2.75, 500x (c); altered coal showing (vac), sample 4, R_o =2.69, 5000x (d).

Approaching the ETS, devolatilization vacuoles (showing clear indications of thermal alteration) become increasingly prevalent in the heat-affected coals (Figures 4(b) and 4(d)). A fine mosaic texture is seen in sample 2 of the number 7 coal seam closest to the intrusion (Figure 4(c)). The heat-affected coal is dominated by vitrinite (83.4–87.6% by volume), with inertinites ranging between 9.8 and 14.2% and the absence of liptinite macerals (Table 1). However, the unaltered coal is dominated by vitrinite (86.4–88.2% by volume), with inertinites ranging between 8.3% and 9.8%. A similar situation was found in the Zhuzhuang mine of the Huaibei Coalfield; the content of vitrinite decreases with the decreasing distance to the intrusions [10]. Coal maceral and coal rank have major effects on the pore size distribution [25, 26].

Proximate Analysis

Moisture, ash, and volatile matter (VM) of the samples are listed in Table 1. The sill intrusion affected coal composition (VM, ash, and moisture) out to a distance of ~160 m. The whole-coal VM (daf basis) decreased from 17.6% (sample 12, ~220 m from the sill) in unaltered coal to 10.0% (sample 1, ~60 m from the sill) in altered coal under the sill (Figure 5(a)).

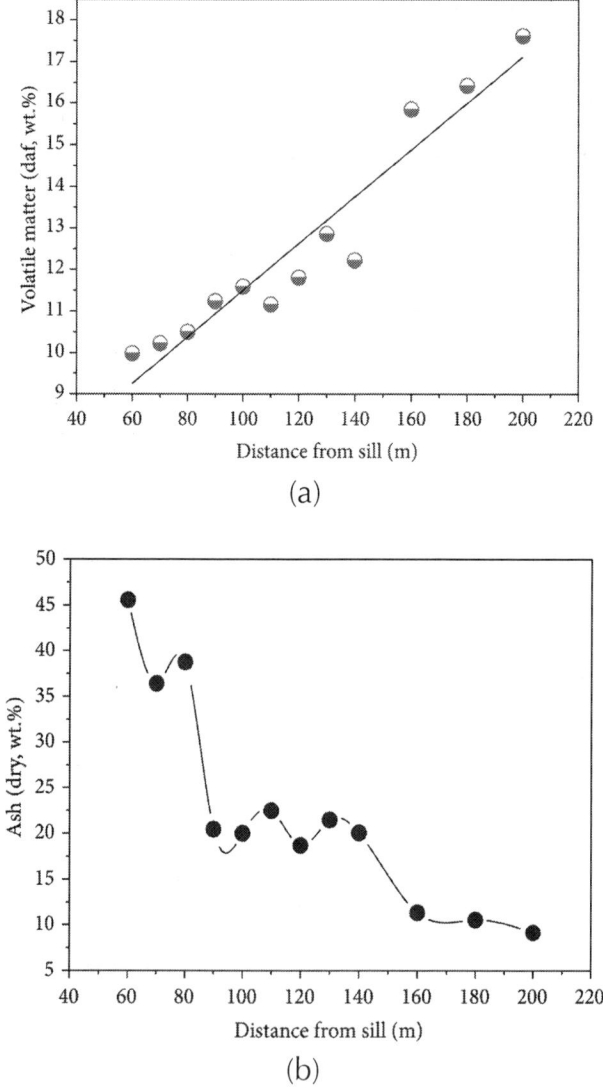

(a)

(b)

Figure 5: Relation between VM (a) and ash (b) of coal and distance from the sill.

The data are grouped in two distinct groups as a function of the distance. The samples 12, 11, and 10 of the number 10 coal seam which are farthest from the intrusion had high values, reaching a maximum of VM = 17.6% (sample 12, ~220 m from the sill), while the samples 1 to 9 belong to another data group, having a minimum of VM = 10.0% (sample 1, ~60 m from the sill). It is concluded that the thermal evolution of sill decreased the VM and increased the coal rank of coal. Accompanying these trends, the moisture decreased from levels of 3.0% to 1.6% under the sill (Table 1). The ash content (dry basis) increased slightly at first from 9.2% (sample 12) to 22.5% (sample 6); then, there was rapid growth to 45.6% (sample 1) in heat-affected coal near the sill (Figure 5(b)). There was higher ash content in coal samples 3 to 1 (belonging to the number 7 coal seam) near the sill intrusion as the result of contact metamorphism. The number 7 coal seam is just under the sill and is the nearest coal seam to the ETS; magma can intrude directly into the number 7 coal seam along the fault fracture zone (Figure 4(a)).

Pore Size Distribution

BET Surface Area

The BET surface areas of the coal samples obtained by using low-pressure nitrogen adsorptive processes are listed in Table 2.

Table 2: Pore characteristics, adsorptive capacity, and initial gas-releasing rates of coal samples

Sample	Units	Number 1	Number 2	Number 3	Number 4	Number 5	Number 6	Number 7	Number 8	Number 9	Number 10	Number 11	Number 12
Distance from sill	(m)	60	70	80	90	100	110	120	130	140	160	180	200
BET surface area	(m²/g)	2.2	2.8	2.8	3.4	3.6	4.2	3.4	3.6	2.4	1.7	1.4	1.5
D-R micr. surface area	(m²/g)	21.9	23.5	25.2	36.4	34.7	35.6	32.5	31.7	33.8	18.1	19.3	17.8
D-A micropore volume	(cm³/g)	0.0079	0.0082	0.0095	0.0146	0.0121	0.0131	0.0138	0.0111	0.0153	0.0088	0.0067	0.0054
Avg. micropore size	(nm)	1.21	1.23	1.24	1.33	1.29	1.31	1.26	1.25	1.27	1.19	1.22	1.18
f		0.42	0.40	0.38	0.22	0.25	0.30	0.29	0.26	0.21	0.17	0.12	0.16
ΔP	(mmHg)	48	43	34	32	34	31	28	31	30	26	21	19
V_l	(m³/t)	37.0	34.2	33.1	35.8	29.9	36.5	31.1	34.7	31.7	26.2	24.6	26.1
P_l	(kPa)	957	808	796	732	1217	822	998	1412	1179	828	1519	812

Note: micropores are pores <2 nm; f: coal firmness coefficient; ΔP: initial gas-releasing rate; V_L: Langmuir volume; P_L: Langmuir pressure (daf basis).

There are obvious differences between samples 1 to 9 and the other samples. Sample 6 (R_o =2.87%, 110 m from the sill) had a BET surface area of 4.2 m²/g, higher than that of other samples. Sample 11 (R_o =2.44%, 180 m from the sill) had a BET surface area of 1.4 m²/g, the lowest value. Approaching the sill, the curve of the BET surface area initially increased and then decreased rapidly close to the ETS (Figure 6). Only for the data of 60–110 m, it seems that the BET surface area decreases close to the intrusion; this may be caused by the coal matrix shrinkage and coal inner heterogeneity when the coal is cooling after the contact metamorphism. The number 7 and 8 coal seams are just under the sill; magma can intrude directly into the coal seams along the fault fracture zone. The trap effect and thermal evolution of the sill obviously increased the BET surface areas of coal (60–160 m from the sill) compared with the coals (160–200 m from the sill).

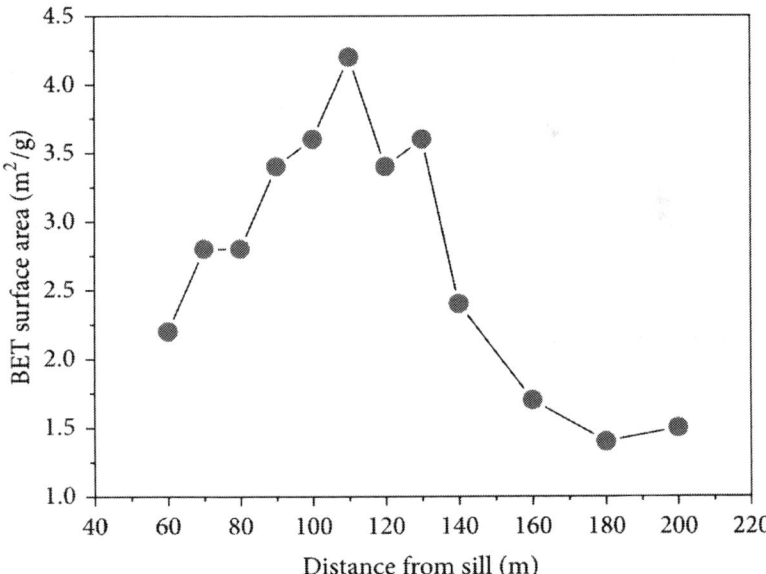

Figure 6: Relation between the BET surface area of coal and distance from sill.

Microporosity Characteristics

The micropore values of coal were obtained using carbon dioxide adsorption, as listed in Table 2. Plotting the Dubinin-Astakhov (D-A) micropore volume and Dubinin-Radushkevitch (D-R) micropore surface area against the distance from the sill, good relativity for the two parameters is shown in Figure 7. The gradients of micropore values are lager near the ETS. This could be due to the contact metamorphism and fluctuations of sill intrusions. In contrast to most coals in other coal-bearing basins in the world, north China coal is well known for its metamorphic complexity resulting from Mesocenozoic igneous intrusions. We simplified the term "multiphasic and superimposed thermal metamorphic evolution" as the fluctuations of igneous intrusions. The uncertainty on microporosity will make the fluctuations of adsorption capacity of Haizi coal.

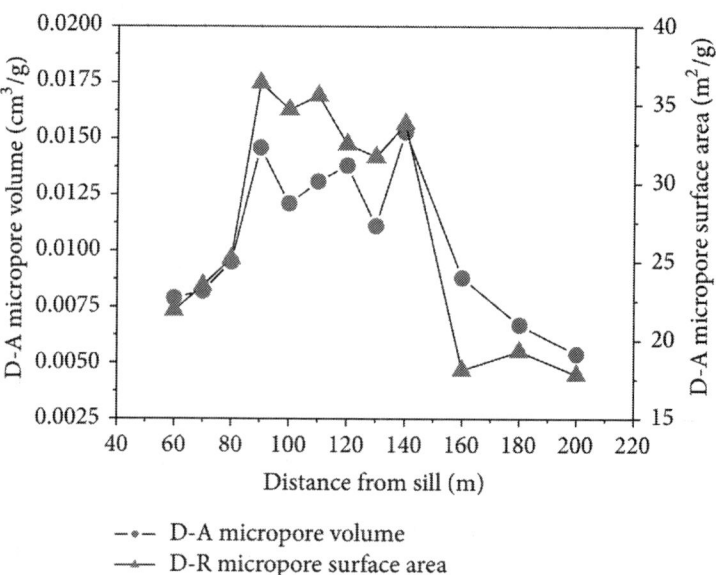

Figure 7: Evolution of micropore volume and surface area with decreasing distance from sill.

The quantities of carbon dioxide adsorbed onto 4 representative samples at relative pressure are shown in Figure 8(a). The pore size distributions as determined by the DA method are shown in Figure 8(b). Note that significantly less CO_2 is adsorbed onto samples 11 and 2 compared with samples 5 and 8. Overall, the micropore volume on account of CO_2 adsorption of the coal appears in the following sequence: coal seam number 8 > coal seam number 9 > coal seam number 7 > coal seam number 10 (Figure 8(a)). The heat-affected coal sample 6 (R_o =2.87%) has the highest average micropore diameter of 1.31 nm, whereas the smallest value of 1.18 nm was found in sample 12 (R_o =2.41%).

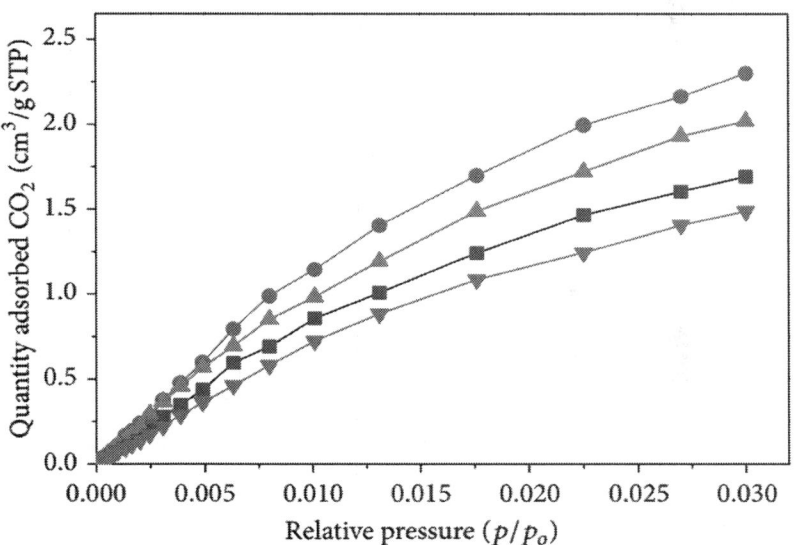

—■— Sample 2, number 7 coal seam
—●— Sample 5, number 8 coal seam
—▲— Sample 8, number 9 coal seam
—▼— Sample 11, number 10 coal seam

(a)

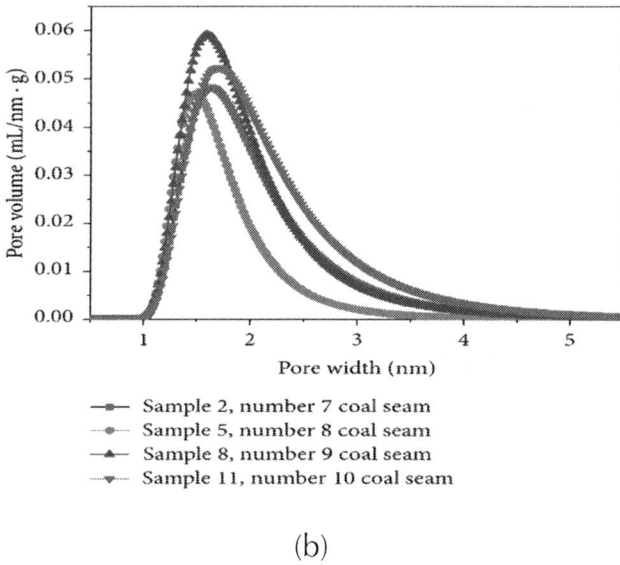

(b)

Figure 8: Quantities of carbon dioxide adsorbed onto coals at relative pressure (a) and microspore size distribution as determined by the DA method (b).

Adsorptive capacity is closely related to micropore (pores < 2 nm) development [25]. Micropore volume is the primary regulator of high-pressure gas adsorption in the Gates coals [27]. The trap effect and thermal evolution of the Yanshan Period sill intrusions obviously increased the micropore volume of coal (60–160 m from the sill) compared with the coals (160–200 m from the sill). This discovery may provide a foundation for understanding the different methane adsorption properties of the Haizi coal samples.

Adsorption and Desorption Properties of Coal

Adsorption Capacity from CH$_4$ Isotherms

The Langmuir volume (V$_L$) and Langmuir pressure (P$_L$) of twelve samples from the adsorption CH$_4$ isotherms are listed in Table 2.

The Langmuir volume of samples initially increased and then decreased approaching the sill, although with fluctuation (Figure 9(a)). Sample 6 (R_o =2.65%, 110 m from the sill) has a maximum V_L of 36.5 m³/t, while sample 11 (R_o =2.32%, 180 m from the sill) has a minimum V_L of 24.6 m³/t. The adsorption capacity of nine altered samples (samples 1 to 9) has an average V_L of 33.0 m³/t which is obviously higher than that of the unaltered coal with an average V_L of 25.6 m³/t. The values of the Langmuir volume are of uncertainty. This could be due to the fluctuations of micropore volumes. However, the uncertainty of data creates difficulties for assessment of coal and gas outburst risk. The methane volume adsorbed of four representative coal sample adsorption isotherms shows a regular pattern; the descending order of the coal seams' adsorptive capacity is numbers 8, 7, 9, and 10 (Figure 9(b)).

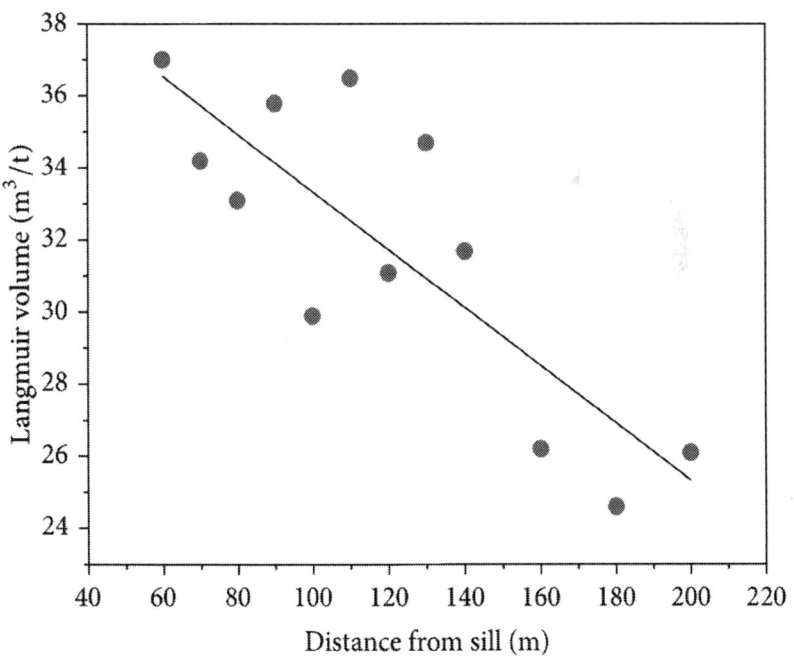

(a)

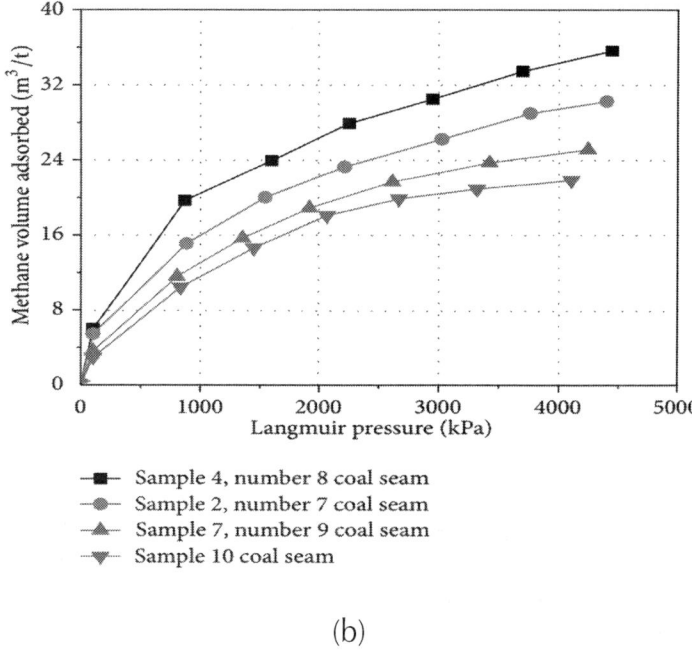

(b)

Figure 9: (a) Evolution of the Langmuir volume with decreasing distance from sill and (b) comparison of gas (methane) adsorption behavior of four samples: sample 2, sample 4, and sample 7 in heat-affected zone and sample 10 in unaltered zone.

The isothermal adsorption curves of three representative altered samples show that sample 2, sample 4, and sample 7, progressively farther from the sill, have a V_L of 34.2, 35.8, and 31.1 m³/t, respectively. These values are much higher than that of the unaltered coal sample 10, sample 11, and sample 12, with an average V_L of 25.6 m³/t (Figure 9). The same regularity (adsorption capacity of altered samples is higher than that of the unaltered coal) was found in the igneous intrusion of Zhuzhuang Mine of the Huaibei Coalfield in previous research [10]. The Langmuir volume has an increased trend with the increasing of micropore volume as determined by the DR method (Figure 10).

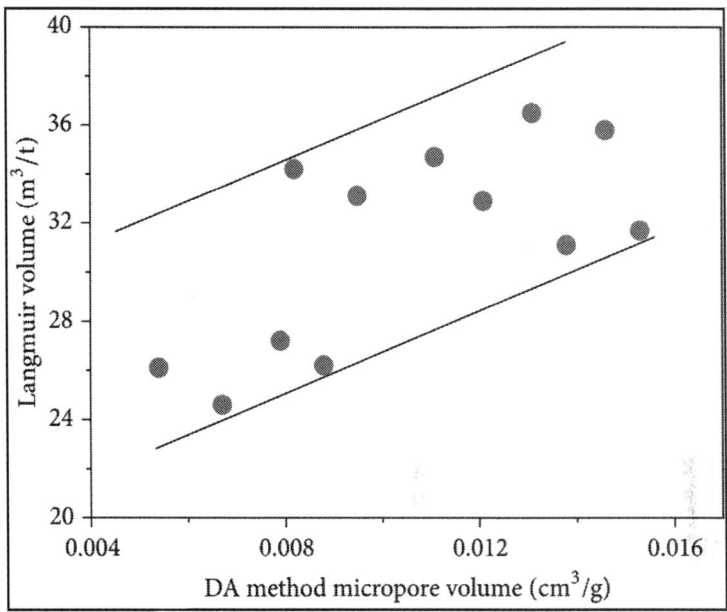

Figure 10: Relationship between the Langmuir volume and the micropore volume as determined by the DA method.

The linear increase of V_L with the micropore volume indicates that microporosity in coal is the governing factor on V_L. The increase of the micropore surface area and the micropore volume enhance the adsorption capacity virtually represented by V_L [22]. Refer to the results in Section 4.1; the thermal evolution of the extremely thick sill obviously increased the coal rank of the Haizi coal under the sill. Coal rank is an important factor affecting the micropore size distribution and thus has an effect on coal adsorption. Approaching the sill, the micropore surface increases with increasing coal rank as does the Langmuir volume.

Methane Initial Gas-Releasing Rate

The ΔP index [28], based on the initial rate of gas desorption from coal, has been widely adopted in Europe and China. In China, the initial gas-releasing rate, ΔP, and the coal firmness coefficient, f, of coal from soft layers are considered two of the four indexes for the identification of coal outbursts. The critical value of ΔP is 10 mmHg, while the critical value of f is 0.5. When the ΔP values > 10 mmHg and $f < 0.5$, the coal is considered prone to outbursts. The coal firmness coefficient reflects the physicomechanical properties of the coal seams. The lower the value is, the more easily the outburst occurs. The ΔP and f measurement results of the twelve coal samples are shown in Table 2. Approaching the sill, both the ΔP and f of the coal samples increased, reaching a maximum of $f = 0.42$ and $\Delta P_{max} = 48$ mmHg, respectively, in sample 1 (Figure 11).

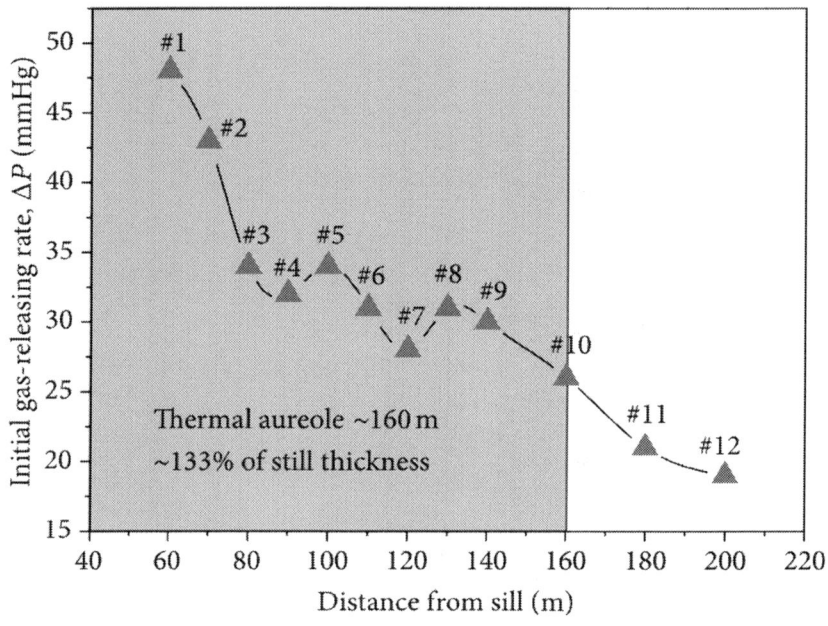

(a)

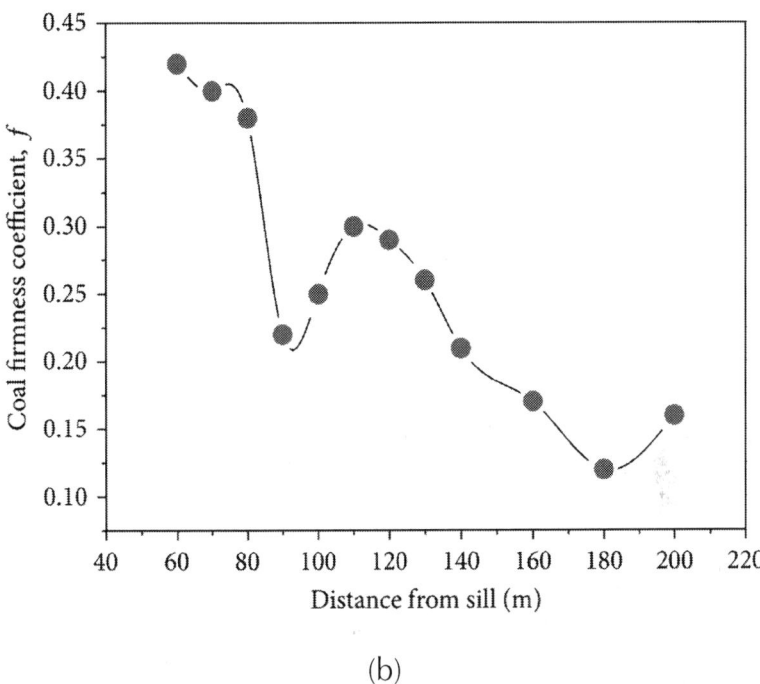

(b)

Figure 11: Relation between methane initial gas-releasing rate ΔP (a), coal firmness coefficient f (b), and distance from sill.

The ΔP values of the 12 samples are larger than the critical value, 10 (Figure 11(a)), while its f values are lower than the critical value, 0.5 (Figure 11(b)). This result verifies that coal seams number 7, 8, 9, and 10 in the Haizi Mine were coal and gas outburst coal seams. The thermal evolution of the Yanshan Period igneous intrusion obviously increased the gas-releasing rate of the coals that were 60–160 m from the sill compared with the coals that were 160–200 m from the sill. In addition, igneous intrusion substantially altered the physicomechanical properties of the surrounding coal mass. The strength of a coaling machine should thus be increased for cutting in a coal seam with frequent igneous intrusions [29]. The hardness of intrusion and of the mineralized Haizi coal adjacent to the sill can require that companies use hard-rock mining equipment to mine through it.

Effects of Sill Intrusion on Coalbed Methane Accumulation

Coalbed methane was rich in the Haizi Mine, with an absolute emission of 52.1 m^3/min and a relative emission of 28.2 m^3/t. A permanent extraction system for coal mine methane exploitation was built in the Haizi Mine, and the extracted methane is used to generate electricity as one of the strategies of greenhouse gas reduction in China [30]. The trap effect and thermal evolution of the sill obviously increased the methane accumulation capacity, the gas pressure, and content of coal seams number 7, 8, and 9 under the ETS. The highest gas pressure observed from number 9 coal seam was 4.50 MPa, with the gas content 17.8 m^3/t. The biggest gas content obtained from number 8 coal seam was 18.3 m^3/t, with a gas pressure 2.80 MPa. The gas pressures of the four coal seams were higher than 0.74 MPa, and the gas contents were higher than 8.00 m^3/t, both higher than the critical values. The gas pressure gradient of number 10 coal seam, in ring-sill trap zone of the Wolonghu Mine, at an elevation above −450 m was 0.003 MPa/m. However, the gradient of elevation below −450 m was 0.027 MPa/m, higher than normal hydrostatic pressure 0.01 MPa/m [15]. This is usually called abnormally high formation pressure [31]. The precise mechanisms of an instantaneous outburst are still unresolved, but the effects of the stress, gas pressure, gas content, and physicomechanical properties of the coal must be considered [32]. Other factors, such as geological features (e.g., igneous intrusions), can combine to exacerbate the problem. The thermal evolution of the Haizi sill obviously increased the capacity of methane generation and retention, while the trap effects of ETS prevented the high-pressure gas from being released. The ETS acted as an impermeable barrier, where pockets of gas have been encountered. Mining activities near the sill would create a low pressure zone, which would allow the adsorptive ability of the enhanced coal to release its gas, leading to localized coalbed methane rapid accumulation and gas outbursts in the Haizi Mine (Figures 1 and 2). Understanding the relationship between outburst disasters and sill intrusions and the mechanism of outbursts, however, requires further rigorous research.

CONCLUSIONS

- The Permian-age coal from the Haizi Mine of the Huaibei Coalfield is intruded by Yanshan Period igneous rock. Thermal alteration occurs down to 160 m (~1.33 × sill thickness). Approaching the extremely thick sill, the values of R_o increased from 2.30% to 2.78%, forming devolatilization vacuoles and a fine mosaic texture. Ash (dry basis) increased from 9.2% to 45.6%. The VM (daf) decreased from 17.6% to 10.0%. The moisture decreased from 3.0% to 1.6%.

- Approaching the sill, the micropore volumes increased from 0.0054 cm^3/g (sample 12, 200 m from the sill) to a maximum 0.0146 cm^3/g (sample 4), followed by a decrease to sample 1 (0.0079 cm^3/g, 60 m from the sill). The BET surface area initially increased and then decreased rapidly close to the ETS. The Langmuir volume increased from 24.6 m^3/t in sample 11 (R_o =2.32%, 180 m from the sill) to 36.5 m^3/t in sample 1. The methane initial gas-releasing rate ΔP increased from 19 mmHg to 48 mmHg.

- The thermal evolution of the Yanshan Period sill intrusions obviously increased the micropore volume, the BET surface area, and methane adsorption capacity of the coal 60–160 m from the sill compared with the coal 160–200 m from the sill. This result may provide a foundation for understanding the different methane adsorption properties of unaltered and heat-affected coals of Haizi coals. The ETS acted as an impermeable barrier. The trap effect of ETS prevented the high-pressure gas from being released, forming gas pockets. Mining activities near the sill would create low pressure zone leading to methane rapid accumulation and gas outbursts in the Haizi Mine.

ACKNOWLEDGMENTS

The authors are grateful to senior engineer Li lei of the Haizi Mine for his assistance with first-hand information on the coal mine. Funding for this research was provided by "A Project Funded by the Priority Academic Program Development of Jiangsu Higher Education Institutions," the Fundamental Research Funds for the Central Universities (no. 2013QNB01), and the Funds for the Key Laboratory of Coalbed Methane Resources and Reservoir Formation Process, Ministry of Education of the CUMT (no. 2013-005).

REFERENCES

1.   J. Cooper, J. Crelling, S. Rimmer, and A. Whittington, "Coal metamorphism by igneous intrusion in the Raton Basin, CO and NM: implications for generation of volatiles," International Journal of Coal Geology, vol. 71, no. 1, pp. 15–27, 2007

2.   L. W. Gurba and C. R. Weber, "Effects of igneous intrusions on coalbed methane potential, Gunnedah Basin, Australia," International Journal of Coal Geology, vol. 46, no. 2–4, pp. 113–131, 2001

3.   Y. B. Yao, D. M. Liu, and W. H. Huang, "Influences of igneous intrusions on coal rank, coal quality and adsorption capacity in Hongyang, Handan and Huaibei coalfields, North China," International Journal of Coal Geology, vol. 88, no. 2-3, pp. 135–146, 2011

4.   L. Wang, Y. P. Cheng, C. Xu, F. H. An, K. Jin, and X. L. Zhang, "The controlling effect of thick-hard igneous rock on pressure relief gas drainage and dynamic disasters in outburst coal seams," Natural Hazards, vol. 66, no. 2, pp. 1221–1241, 2013

5.   S. F. Dai and D. Y. Ren, "Effects of magmatic intrusion on mineralogy and geochemistry of coals from the Fengfeng-Handan coalfield, Hebei, China," Energy & Fuels, vol. 21, no. 3, pp. 1663–1673, 2007

6. M. Mastalerz, A. Drobniak, and A. Schimmelmann, "Changes in optical properties, chemistry, and micropore and mesopore characteristics of bituminous coal at the contact with dikes in the Illinois Basin," International Journal of Coal Geology, vol. 77, no. 3-4, pp. 310–319, 2009

7. S. M. Rimmer, L. E. Yoksoulian, and J. C. Hower, "Anatomy of an intruded coal, I: effect of contact metamorphism on whole-coal geochemistry, Springfield (No. 5) (Pennsylvanian) coal, Illinois Basin,"International Journal of Coal Geology, vol. 79, no. 3, pp. 74–82, 2009

8. A. Schimmelmann, M. Mastalerz, L. Gao, P. E. Sauer, and K. Topalov, "Dike intrusions into bituminous coal, Illinois Basin: H, C, N, O isotopic responses to rapid and brief heating," Geochimica et Cosmochimica Acta, vol. 73, no. 20, pp. 6264–6281, 2009

9. S. Sarana and R. Kar, "Effect of igneous intrusive on coal microconstituents: study from an Indian Gondwana coalfield," International Journal of Coal Geology, vol. 85, no. 1, pp. 161–167, 2011

10. Y. B. Yao and D. M. Liu, "Effects of igneous intrusions on coal petrology, pore-fracture and coalbed methane characteristics in Hongyang, Handan and Huaibei coalfields, North China," International Journal of Coal Geology, vol. 96–97, pp. 72–81, 2012.

11. A. N. Golab and P. F. Carr, "Changes in geochemistry and mineralogy of thermally altered coal, Upper Hunter Valley, Australia," International Journal of Coal Geology, vol. 57, no. 3-4, pp. 197–210, 2004.

12. A. Saghafi, K. Pinetown, P. Grobler, and J. Vanheerden, "CO_2 storage potential of South African coals and gas entrapment enhancement due to igneous intrusions," International Journal of Coal Geology, vol. 73, no. 1, pp. 74–87, 2008.

13. X. Q. He, W. X. Chen, B. S. Nie, and M. Zhang, "Classification technique for danger classes of coal and gas outburst in deep coal mines," Safety Science, vol. 48, no. 2, pp. 173–178, 2010.

14. S. B. Anderson, "Outbursts of methane gas and associated mining problems experienced at Twistdraai Colliery," in Proceedings of the International Symposium Cum Workshop on Management & Control of High Gas Emissions & Outbursts, R. Lama, Ed., pp. 423–434, Wollongong, Australia, 1995.

15. J. Y. Jiang, Y. P. Cheng, L. Wang, W. Li, and L. Wang, "Petrographic and geochemical effects of sill intrusions on coal and their implications for gas outbursts in the Wolonghu Mine, Huaibei Coalfield, China," International Journal of Coal Geology, vol. 88, no. 1, pp. 55–66, 2011.

16. L. G. Zheng, G. J. Liu, L. Wang, and C. L. Chou, "Composition and quality of coals in the Huaibei Coalfield, Anhui, China," Journal of Geochemical Exploration, vol. 97, no. 2-3, pp. 59–68, 2008.

17. D. M. Liu, Y. B. Yao, D. Z. Tang, S. H. Tang, C. Yao, and W. H. Huang, "Coal reservoir characteristics and coalbed methane resource assessment in Huainan and Huaibei coalfields, Southern North China,"International Journal of Coal Geology, vol. 79, no. 3, pp. 97–112, 2009.

18. S. F. Han, Coal-Forming Geological Conditions and Forecast of Huainan and Huaibei Coalfield, Geology, Beijing, China, 1990.

19. ASTM, "Section five, petroleum products, lubricants, and fossil fuels," in Annual Book of ASTM Standards, vol. 05.06, ASTM International, West Conshohocken, Pa, USA, 2007.

20. "Methods for the petrographic analysis of bituminous coal and anthracite—part 5: method of determining microscopically the reflectance of vitrinite," ISO 7404-5, 1994.

21. G. H. Taylor, M. Teichmüller, A. Davis, C. F. K. Diessel, R. Littke, and P. Robert, Organic Petrology, Gebrüder Borntraeger, Berlin, Germany, 1984.

22. F. H. An, Y. P. Cheng, D. M. Wu, and L. Wang, "The effect of small micropores on methane adsorption of coals from Northern China," Journal of the International Adsorption Society, vol. 19, no. 1, pp. 83–90, 2013.

23. "State Administration of Coal Mine Safety of China," Determination Method for Index (ΔP) of Initial Velocity of Diffusion of Coal Gas, 2009.

24. C. E. Barker and M. J. Pawlewicz, "Calculation of vitrinite reflectance from thermal histories and peak temperatures: a comparison of methods," in Vitrinite Reflectance as a Maturity Parameter, P. K. Mukhopadhyay and W. G. Dow, Eds., ACS Symposium, pp. 216–229, American Chemical Society, 1994.

25. P. J. Crosdale, B. B. Beamish, and M. Valix, "Coalbed methane sorption related to coal composition," International Journal of Coal Geology, vol. 35, no. 1–4, pp. 147–158, 1998.

26. M. Mastalerz, A. Drobniak, and J. Rupp, "Meso-and micropore characteristics of coal lithotypes: implications for CO_2 adsorption," Energy & Fuels, vol. 22, no. 6, pp. 4049–4061, 2008.

27. C. R. Clarkson and R. M. Bustin, "The effect of pore structure and gas pressure upon the transport properties of coal: a laboratory and modeling study 1. Isotherms and pore volume distributions," Fuel, vol. 78, no. 11, pp. 1333–1344, 1999.

28. I. L. Ettinger, E. S. Zhupakhina, and L. E. Schterenberg, "Methods of Allowing Forecasting in the Seams of Coal Zoned Subject to Instantaneous Outbursts," Extract from Academy of Sciences of the USSR, Institute of Mines Central Committee of Measures Against Instantaneous Outbursts, Moscow, Cherchar Translation No. 111-50, Paris, France, 1958.

29. R. Singh, A. K. Singh, and P. K. Mandal, "Cuttability of coal seams with igneous intrusions," Engineering Geology, vol. 67, no. 1-2, pp. 127–137, 2002.

30. Y. P. Cheng, L. Wang, and X. L. Zhang, "Environmental impact of coal mine methane emissions and responding strategies in China," International Journal of Greenhouse Gas Control, vol. 5, no. 1, pp. 157–166, 2011.

31. X. Su, X. Lin, S. Liu, M. Zhao, and Y. Song, "Geology of coalbed methane reservoirs in the Southeast Qinshui Basin of

China," International Journal of Coal Geology, vol. 62, no. 4, pp. 197–210, 2005.

32.   B. B. Beamish and P. J. Crosdale, "Instantaneous outbursts in underground coal mines: an overview and association with coal type," International Journal of Coal Geology, vol. 35, no. 1–4, pp. 27–55, 1998.

Chapter 2

Evaluation of Multiple-Scale 3D Characterization for Coal Physical Structure with DCM Method and Synchrotron X-Ray CT

Haipeng Wang[1], Yushuang Yang[1, 2], Jianli Yang[3], Yihang Nie[1], Jing Jia[4], and Yudan Wang[5]

[1]Institute of Theoretical Physics and Department of Physics, Shanxi University, Taiyuan, Shanxi 030006, China

[2]CSIRO, Clayton, VIC 3169, Australia

[3]State Key Laboratory of Coal Conversion, Institute of Coal Chemistry, Chinese Academy of Sciences, Taiyuan, Shanxi 030001, China

[4]College of Physics & Electronics Engineering, Shanxi University, Taiyuan, Shanxi 030006, China

[5]Shanghai Institute of Applied Physics, Chinese Academy of Sciences, Shanghai 201800, China

ABSTRACT

Multiscale nondestructive characterization of coal microscopic physical structure can provide important information for coal conversion and coal-bed methane extraction. In this study, the physical structure of a coal sample was investigated by synchrotron-based multiple-energy X-ray CT at three beam energies and two different spatial resolutions. A data-constrained modeling (DCM) approach was used to quantitatively characterize the multiscale compositional distributions at the two resolutions. The volume fractions of each voxel for four different composition groups were obtained at the two resolutions. Between the two resolutions, the difference for DCM computed volume fractions of coal matrix and pores is less than 0.3%, and the difference for mineral composition groups is less than 0.17%. This demonstrates that the DCM approach can account for compositions beyond the X-ray CT imaging resolution with adequate accuracy. By using DCM, it is possible to characterize a relatively large coal sample at a relatively low spatial resolution with minimal loss of the effect due to subpixel fine length scale structures.

INTRODUCTION

Physically, a coal sample can be grouped in three compositions: coal matrix (organic composition), minerals (inorganic compositions), and pores. The physical structure of coal is relevant to the sorption and diffusion of coal-bed methane (CBM) in coal-bed as well as the transformation of minerals in coal processing [1–7]. The distributions of coal compositions (physical structures) are inherently heterogeneous and multiscale, ranging from nanometer to millimeter scale and above. Multiple-scale three-dimensional characterization of coal physical structure is helpful for clean and high efficient utilization of coal. It is also useful for obtaining fundamental data to establish three-dimensional (3D) fluid transportation model in coal matrix during enhanced coal-bed methane (ECBM) process.

X-ray CT imaging can nondestructively obtain 3D materials microstructure information. Numerous works related to the application of X-ray CT in coal microstructure characterization have been done previously. Verhelst et al. [8] and Simons et al. [9] investigated the correlation between the tomodensity with the real physical bulk density of coal compositions. Van Geet et al. [10, 11] demonstrated the application of microfocus CT and the use of dual energy approach. Karacan and Okandan [12, 13], Mazumder et al. [14], and Yao et al. [15, 16] also studied the distribution of different compositions in coal sample by X-ray CT imaging. At present, microfocus CT combined with dual-energy method and image segmentation has been an important method for quantitative characterization of the physical structure of coal.

Although remarkable progresses have been made by previous researchers, there are still some limitations with this technique in coal characterization. Firstly, in most cases the characterization of the coal physical structure is a multiscale problem. For example, pore sizes ranging from several hundred micrometers to several nanometers all contribute to coal-bed methane transportation. However, there exists a scale limitation with existing 3D characterization techniques such as X-ray CT: it is not possible for the sample size to go beyond 10^4 times of the scanned voxel size. That is to say, for a sample with a size of centimeters, the imaging resolution size could not be smaller than a micrometer. For a factor 10 increase in resolution, the CT data size is increased by 10^3 times. In addition to this, the X-ray exposure increases as the 4th power with resolution or 10^4 times with a factor 10 increase in resolution. The dataset size, exposure induced sample temperature increase, and image acquisition time would become unpractical for a moderate further improvement in resolution. In order to obtain multiple-scale information of various materials microstructure, micro-CT was usually combined with other high resolution techniques during the characterization process, such as the combination of micro-CT and nanofocus CT [17], scanning electron microscope (SEM) [18], and focused ion beam (FIB) [19]. Despite the successes, the high resolution techniques can only provide details for a small region of a sample or a small sample, while for some nonhomogenous materials a large sample would be more accurate representation statically. Secondly, the fine length scale structures in the coal sample produce effective mixing of multiple compositions at the X-ray CT voxels. This leads to a nonunique relationship between material compositions and X-ray CT image grey scale. This makes it inadequate to use the conventional image segmentation technique to resolve the compositions.

Recently, a data-constrained modeling (DCM) approach [20, 21] has been developed. It combines the multispectrum X-ray CT data with a statistical mechanical model to resolve the coexistence of multiple compositions in the same voxel (the partial volume effect). It has been successfully used to characterize microstructures of different materials.

Wang et al. [22] have applied this approach to characterize an anthracite coal sample collected from Yangquan coal mine, China. The mineral compositions, coal matrix, and pores were divided into four groups based on their X-ray absorption characteristics. The volume fractions of each composition group in individual CT voxel were obtained using the DCM approach. Fine compositional structures which are smaller than the CT imaging resolution contribute to voxel partial volumes. Through using DCM, the compositions smaller than the X-ray CT spatial resolution can be characterized quantitatively. This method has opened new opportunity for multiple-scale 3D characterization of coal.

Despite the success of previous work, the sensitivity of the results on imaging spatial resolution is still unknown. The purpose of this study is to establish the spatial resolution sensitivity for coal microstructure characterization using DCM and multiple X-ray CT datasets acquired at different spatial resolution. It forms the basis for multiple-scale characterization of coal samples.

EXPERIMENTAL

Sample and CT Experiment

The coal sample used in this study is the same one as used by previous study and the mineral compositions, coal matrix, and pores were divided into four groups (as listed in Table 1) based on their X-ray absorption characteristics [22]. That is, those compositions were treated as one group when their X-ray absorption coefficients as functions of X-ray energy were approximately linearly dependent on each other.

Table 1: Compositions groups of coal sample

Groups	A	B	C	D
Including compositions	Pore	Coalmatrix	Illite, quartz, and kaolinite	Chlorite and titania

For convenience of image alignment with different resolutions, a short nylon wire was stuck to the sample (Figure 1) as a marker. There was no obvious pore on the sample surface.

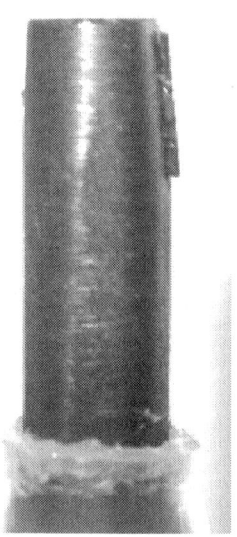

Figure 1: Coal sample used in this study.

The X-ray CT projection data was acquired on the BL13W beamline at the Shanghai Synchrotron Radiation Facility (SSRF). A Si (111) double-crystal monochromator was used, which produces a quasi-monochromatic X-ray beam with a relative bandwidth smaller than 5×10^{-3}. The excellent monochromaticity of synchrotron X-ray makes the correlation between the tomodensity with the physical density of the coal sample more easy to be established. An Optique Peter X-ray CCD detector with a native pixel size of $7.4 \times 7.4\,\mu m^2$ was used in the experiment. With a 2× optical lens in front of the CCD, the sample was imaged with an imaging resolution of $3.7\,\mu m$. Monochromatic X-ray beam energies of 14 kev, 18 kev, and 22 kev were selected. A total of 900 projection images were collected at each X-ray beam energy. The angular spacing was 0.2° between each two projections for a total rotation angle of 180°. After this, the sample was imaged again with an imaging resolution of $11.84\,\mu m$ using a 1.25× optical lens and 2×2 pixels binning. The position of sample and the orientation of the sample axis were not changed when the optical lens was changed. This can ensure the same portion of the sample was imaged at the abovementioned two imaging resolutions. For imaging resolution of $11.84\,\mu m$, the same X-ray beam energies were used as for imaging resolution of 3.7 μm. A total of 360 projection images were collected at each X-ray beam energy. The projection images were acquired with an angular spacing 0.5° for a total sample rotation angle of 180°. The dark-current and flat-field images were also acquired at the beginning and at the end of each scan which are used to normalize the projection images before CT reconstruction. For imaging resolution size of 3.7 μm, a total of 1395 image slices with the size of 1947 × 1947 pixels were reconstructed using the X-TRACT software [23] at each beam-energy. For image resolution size of 11.84 μm, a total of 697 image slices with the size of 973 × 973 pixels were reconstructed at each beam energy.

DCM Approach

The sample was represented by a cubic grid of N = l × l × n voxels in the DCM model, where l × l is the total pixel size of a CT slice and n is the number of selected slices. The voxel size in the DCM model is identical to the voxel size of reconstructed X-ray CT slices. The DCM model is to minimize the following objective function at each voxel [20, 22]:

$$\Gamma = \sum_{q=1}^{3} \left[\sum_{m=0}^{3} \mu^{(q,m)} v^{(m)} - \hat{\mu}^{(q)} \right]^2 + \sum_{m=0}^{3} v^{(m)} \varepsilon^{(m)}.$$

(1)

In (1), the q = (= 1, 2, 3) corresponds to the experimental CT data at beam energies 14, 18, and 22 keVs, respectively; m(= 0, 1, 2, 3) represent the compositional groups A, B, C, and D, respectively; $v^{(m)}$ is volume fractions of composition group m on the voxel; $\mu^{(q, m)}$ is the linear absorption coefficients of composition group m at the X-ray energy q; $\varepsilon^{(m)}$ is the phenomenological chemical potentials of composition group m; $\hat{\mu}$ (q= 1, 2, 3) is X-ray CT measured linear absorption coefficient values on the voxel. The following constraints should be applied in minimization of (1):

$$0 \leq v^{(0)}, v^{(1)}, v^{(2)}, v^{(3)} \leq 1$$

$$v^{(0)} + v^{(1)} + v^{(2)} + v^{(3)} = 1.$$

(2)

For the DCM approach, the volume fractions of each of the composition groups can be obtained by solving (1) and (2) using a constrained search algorithm.

For the purpose of comparing the DCM reconstructed compositions at different resolutions, a subset of 96 slices with pixel size of 3.7 μm and a subset of 30 slices with pixel size of 11.84 μm were selected at each X-ray beam energy. The selected slices with different voxel sizes correspond to the same physical portion of the coal sample. The subset was selected with minimal image defects. The composition distributions at the two resolutions were computed with the DCM approach. For quantitative comparisons of DCM reconstructed compositions in different regions, seven cubical subregions of the selected slices were used. The size of the subregions is so selected such that each of them contains an integer number of voxels at both resolutions. This ensures that the subregions at different resolutions are corresponding to the same portion of the coal sample. The size of a subregion is 96 × 96 × 96 pixels at a resolution 3.7 μm and 30 × 30 × 30 pixels at a resolution 11.84 μm, respectively. During the DCM computation process, the pore self-energy was selected as 0.0000805 to reproduce the previous measured porosity of the coal sample, and other parameters were the same as previous study [22].

RESULTS AND DISCUSSION

Figure 2 shows two typical reconstructed CT slices at spatial resolution of 3.7 μm and 11.84 μm respectively. They correspond to the same sample slice. A voxel in Figure 2(b) is equivalent to 3.2^3 voxels in Figure 2(a). The X-ray linear absorption coefficient of a voxel in Figure 2(b) is the average of those 3.2^3 voxels in Figure 2(a). There are obvious ring artifacts existing in the images, which are difficult to be eliminated completely during the CT reconstruction, although a ring-defect filter had been applied. In these two images, the grey areas are mainly coal matrix; the white areas indicate high X-ray absorption, which represent minerals. From Figure 2, it is easy to see that although the two slices ((a) and (b)) look similar, the details are clearer in the image with the higher resolution. The reason for this is that Figure 2(a) contained more pixels than Figure 2(b), which enable a more detailed characterization for the sample.

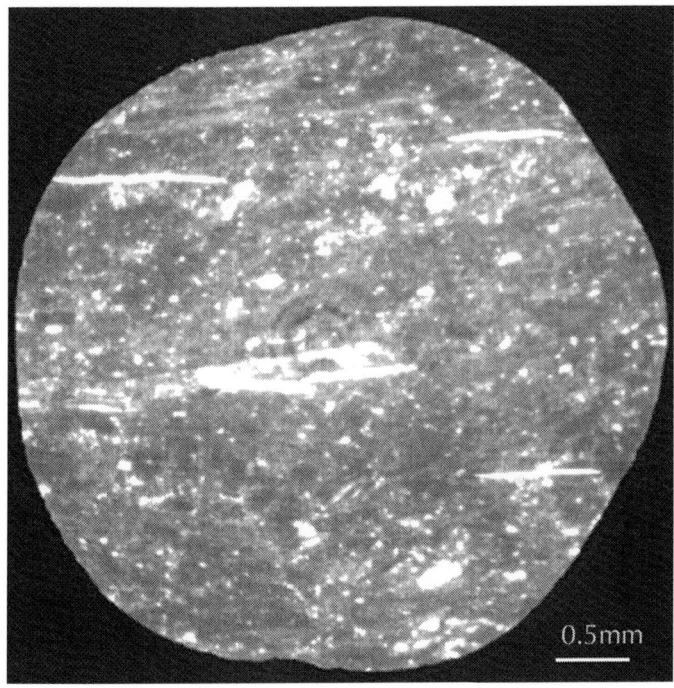

0.5mm

(a)

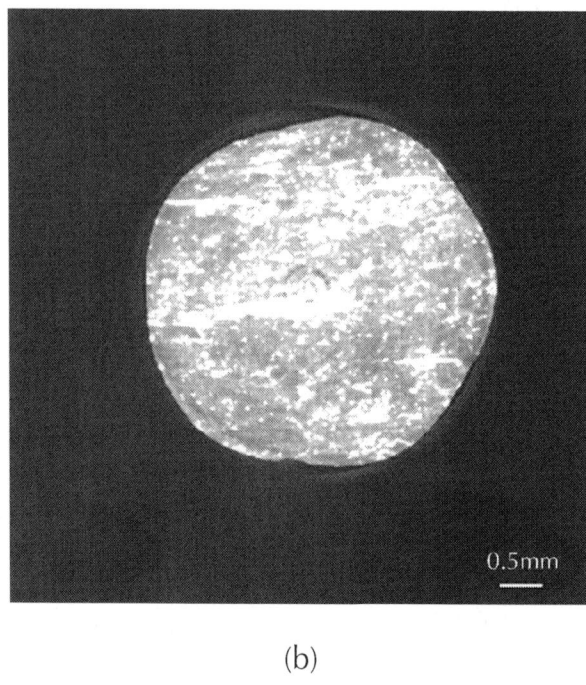

(b)

Figure 2: The original reconstructed CT slices at different spatial resolutions: (a) at resolution of 3.7 μm; (b) at resolution of 11.84 μm.

Table 2 is the DCM computed average composition volume fractions at the two CT imaging resolutions. In Table 2, Area 0 represents DCM computed results of the whole selected subset slices of the coal sample. Areas 1–7 represent DCM computed results of the seven subregions.

Table 2: DCM computed average composition group volume fractions

Area	Imaging resolution (μm)	Computed volume fractions (%)				Computed time (s)
		Group A	Group B	Group C	Group D	
Area 0	3.7	4.48	87.80	7.58	0.14	9.06×10^5
	11.84	4.37	88.00	7.51	0.12	6.55×10^4
Area 1	3.7	3.30	86.90	9.40	0.40	2447
	11.84	3.39	86.90	9.31	0.40	70

Area 2	3.7	5.25	89.30	5.40	0.05	2428
	11.84	4.97	89.60	5.36	0.07	79
Area 3	3.7	3.27	85.40	11.10	0.23	2274
	11.84	3.36	85.40	11.00	0.24	71
Area 4	3.7	4.82	89.10	6.02	0.06	2366
	11.84	4.63	89.40	5.85	0.12	72
Area 5	3.7	4.14	88.50	7.22	0.14	2547
	11.84	4.25	88.20	7.37	0.18	70
Area 6	3.7	4.35	88.20	7.28	0.17	2415
	11.84	4.21	88.40	7.16	0.23	73
Area 7	3.7	4.52	89.20	6.17	0.11	2386
	11.84	4.41	89.40	6.07	0.12	70

It can be seen from Table 2 that the computed volume fractions of different composition groups for the two image resolutions are very close. The difference of the computed volume fractions for coal matrix and pore is less than 0.3%. For the mineral compositions group, the difference is less than 0.17%. It should be note that though the maximal difference for pore in the eight areas is 0.28%, the average difference is only 0.14%. The minor differences may be related to the experimental noise during the projection image acquisition process and the ring artifact in the CT slices. The difference of coal matrix is higher than other compositions, since the volume fraction of coal matrix in coal is far more than other compositions. The results indicate that the DCM approach can be used to characterize coal microstructures using low resolution CT data with minimal loss of fine subvoxel structure effects. For the same physical volume in the coal sample, about 30 times more computational time is required for the image resolution at 3.7 μm than that at 11.84 μm as shown in Table 2.

Figures 3(a) and 3(b) show the DCM computed compositional distributions at two different image resolutions. They are on the same portion of the sample as in Figures 2(a) and 2(b). Although Figures 3(a) and 3(b) are quite similar overall, the composition of group D (green area) is more visible in Figure 3(a) than that in Figure 3(b). The reason for this is that the display intensity of fine particles at a coarse display resolution is suppressed as their small volume fractions tend to be overwhelmed by other compositions. In other words, the small regions of group D appear as diluted and spread out in Figure 3(b).

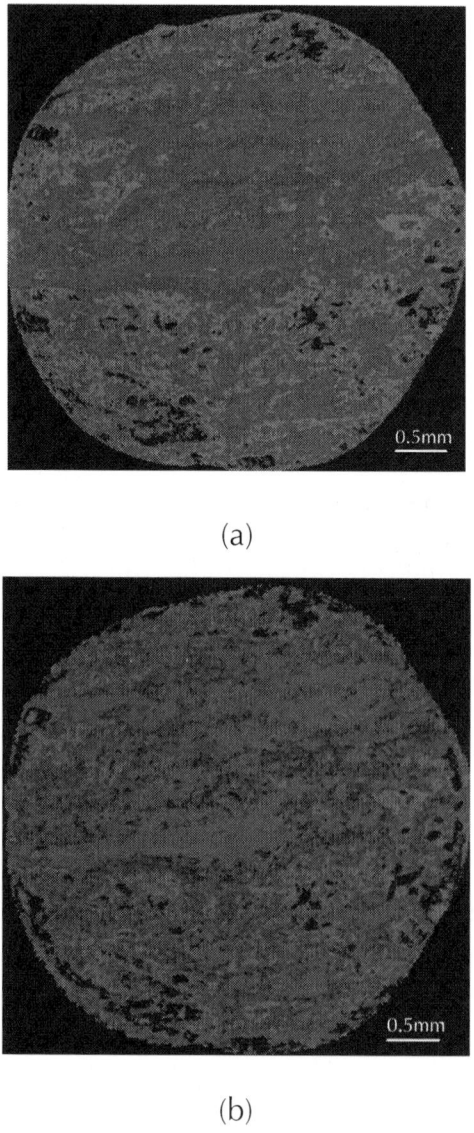

(a)

(b)

Figure 3: DCM reconstructed compositional distributions: (a) at resolution of 3.7 μm; (b) at resolution of 11.84 μm. Pores are displayed as red;

group C minerals are displayed as blue; group D minerals are displayed as green. The pixel display intensity for each colour is proportional to the appropriate compositional volume fraction. Coexistence of multiple compositions in a voxel is shown as colour mixing. The coal matrix is not displayed.

Figures 4(a) and 4(b) are the 3D physical structures of the coal sample corresponding to Area 1 at different spatial resolutions. It can be seen from these two images that the pores are distributed as a 3D connected network, while group C mineral composition is distribution as clusters. The apparent intensity of red color (pore) in Figure 4(a) is higher than that in Figure 4(b), for the same reason as has been mentioned in Figure 3. The visual effect of such dispersed compositions can be enhanced by multiplying their volume fractions values by a suitable factor (>1).

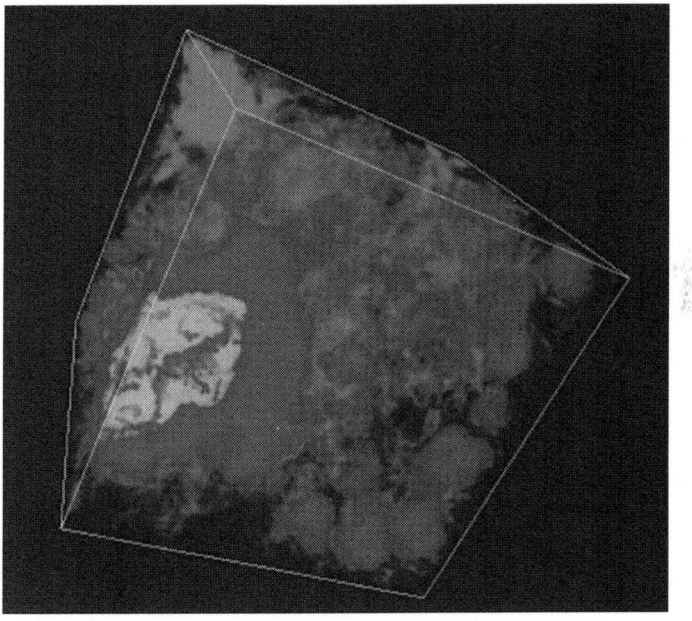

(a)

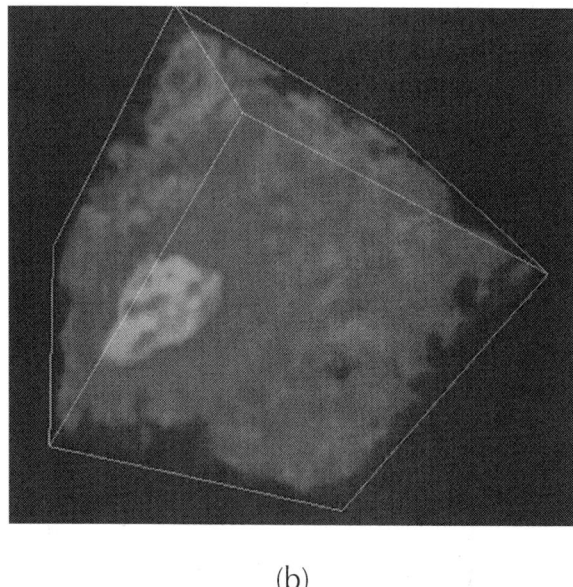

(b)

Figure 4: 3D physical structure of Area 1 at different spatial resolutions: (a) at resolution of 3.7 μm; (b) at resolution of 11.84 μm. The same colouring scheme is used as in Figure 3. The displayed sample size is 355.2 × 355.2 × 355.2 μm³.

Although the numerical results indicate that DCM can model the partial volume effect reliably, it should be noted that it does not increase the spatial resolution. That is, it does not give any additional information about how the multiple compositions are spatially distributed inside a voxel. Investigations on the impact of subvoxel spatial distributions still need high resolution techniques. Besides this, the DCM still cannot to be used for lab-CT. The X-ray of lab-CT is polychromatic, which make the correlation between the tomodensity with the physical density of sample more difficult to be established.

CONCLUSIONS

The 3D physical structure of a Yangquan coal sample is obtained using the DCM nonlinear optimization approach with multienergy synchrotron X-ray CT at different image resolutions. The distributions of pores, coal matrix, and minerals in the same region of the coal sample have been quantitatively reconstructed at different spatial resolutions. By comparing average volume fractions of mineral phases and pores in these subregions in the sample, it is concluded that the DCM computed results were consistent under the two resolutions. For materials with structures span over multiple length scales such as coal, it has been difficult to characterize them using the existing 3D X-ray CT technique due to limited spatial dynamical range of the technique. If one wish to resolve the fine length scale structures, the sample size has to be reduced accordingly that would reduce its statistical representativeness. The results presented in this paper indicate that using a large sample would not necessarily lose subvoxel effect by using the DCM approach, although, with higher image resolution, spatial distributions of compositions whose size is smaller than the pixel size can be resolved more precisely. With a reduced spatial resolution, the computational efficiency is improved.

CONFLICT OF INTERESTS

The authors declare that they have no financial or personal relationships with other people or organizations that can inappropriately influence their works; there are no professional or other personal interests of any nature or kind in any product, service, and/or company that could be construed as influencing the position presented in, or the review of, the paper.

ACKNOWLEDGMENTS

This work is sponsored by the National Key Basic Research Program of China (973) (no. 2014CB239004), National Nature Science Foundation of China (no. 21206087), the One Hundred Person Project of Shanxi Province, and CSIRO Computational and Simulation Science Transformational Capability Platform. The authors would like to thank BL13W of the Shanghai Synchrotron Radiation Facility for providing X-ray CT experimental support.

REFERENCES

1.   A. D. Alexeev, E. P. Feldman, and T. A. Vasilenko, "Kinetics of methane desorption from coal nano-and mesostructures," Energy and Fuels, vol. 24, no. 8, pp. 4375–4379, 2010.

2.   M. Pillalamarry, S. Harpalani, and S. Liu, "Gas diffusion behavior of coal and its impact on production from coalbed methane reservoirs," International Journal of Coal Geology, vol. 86, no. 4, pp. 342–348, 2011.

3.   F. S. Han, A. Busch, B. M. Krooss, Z. Y. Liu, and J. L. Yang, "CH4 and CO2 sorption isotherms and kinetics for different size fractions of two coals," Fuel, vol. 108, pp. 137–142, 2013.

4.   R. C. Everson, H. W. J. P. Neomagus, and R. Kaitano, "The random pore model with intraparticle diffusion for the description of combustion of char particles derived from mineral- and inertinite rich coal," Fuel, vol. 90, no. 7, pp. 2347–2352, 2011.

5.   S. Hol, C. J. Spiers, and C. J. Peach, "Microfracturing of coal due to interaction with CO2 under unconfined conditions," Fuel, vol. 97, pp. 569–584, 2012.

6.   W. N. Yuan, Z. J. Pan, X. Li et al., "Experimental study and modelling of methane adsorption and diffusion in shale," Fuel, vol. 117, pp. 509–519, 2014.

7.   A. Golab, C. R. Ward, A. Permana, P. Lennox, and P. Botha, "High-resolution three-dimensional imaging of coal using microfocus X-ray computed tomography, with special reference to modes of mineral occurrence," International Journal of Coal Geology, vol. 113, pp. 97–108, 2013.

8.   F. Verhelst, P. David, W. Fermont, L. Jegers, and A. Vervoort, "Correlation of 3D-computerized tomographic scans and 2Dcolour image analysis of Westphalian coal by means of

multivariate statistics," International Journal of Coal Geology, vol. 29, no. 1–3, pp. 1–21, 1996.

9.   F. J. Simons, F. Verhelst, and R. Swennen, "Quantitative characterization of coal by means of microfocal X-ray computed microtomography (CMT) and color image analysis (CIA)," International Journal of Coal Geology, vol. 34, no. 1-2, pp. 69– 88, 1997.

10.   M. Van Geet, R. Swennen, and M. Wevers, "Quantitative analysis of reservoir rocks by microfocus X-ray computerised tomography," Sedimentary Geology, vol. 132, no. 1-2, pp. 25–36, 2000.

11.   M. Van Geet, R. Swennen, and P. David, "Quantitative coal characterisation by means of microfocus X-ray computer tomography, colour image analysis and back-scattered scanning electron microscopy," International Journal of Coal Geology, vol. 46, no. 1, pp. 11–25, 2001.

12.   C. O. Karacan and E. Okandan, "Fracture/cleat analysis of coals ¨ from Zonguldak Basin (northwestern Turkey) relative to the potential of coalbed methane production," International Journal of Coal Geology, vol. 44, no. 2, pp. 109–125, 2000.

13.   C. O. Karacan and E. Okandan, "Adsorption and gas transport in coal microstructure: investigation and evaluation by quantitative X-ray CT imaging," Fuel, vol. 80, no. 4, pp. 509– 520, 2001.

14.   S. Mazumder, K.-H. A. A. Wolf, K. Elewaut, and R. Ephraim, "Application of X-ray computed tomography for analyzing cleat spacing and cleat aperture in coal samples," International Journal of Coal Geology, vol. 68, no. 3-4, pp. 205–222, 2006.

15.   Y. B. Yao, D. M. Liu, Y. Che, D. Z. Tang, S. S. Tang, and W. H. Huang, "Non-destructive characterization of coal samples from China using microfocus X-ray computed tomography," International Journal of Coal Geology, vol. 80, no. 2, pp. 113–123, 2009.

16.   Y. B. Yao, D. M. Liu, Y. D. Cai, and J. Q. Li, "Advanced characterization of pores and fractures in coals by nuclear

magnetic resonance and X-ray computed tomography," Science China Earth Sciences, vol. 53, no. 6, pp. 854–862, 2010.

17.   B. M. Patterson, K. C. Henderson, P. J. Gibbs, S. D. Imhoff, and A. J. Clarke, "Laboratory micro- and nanoscale X-ray tomographic investigation of Al–7at.%Cu solidification structures," Materials Characterization, vol. 95, pp. 18–26, 2014.

18.   R. Quey, H. Suhonen, J. Laurencin, P. Cloetens, and P. Bleuet, "Direct comparison between X-ray nanotomography and scanning electron microscopy for the microstructure characterization of a solid oxide fuel cell anode," Materials Characterization, vol. 78, pp. 87–95, 2013.

19.   K. E. Yazzie, J. J. Williams, N. C. Phillips, F. de Carlo, and N. Chawla, "Multiscale microstructural characterization of Sn-rich alloys by three dimensional (3D) X-ray synchrotron tomography and focused ion beam (FIB) tomography," Materials Characterization, vol. 70, pp. 33–41, 2012.

20.   Y. S. Yang, "A data-constrained non-linear optimisation approach to compositional microstructure prediction," Lecture Notes in Information Technology, vol. 15, pp. 198–205, 2012.

21.   S. Yang, A. Tulloh, F. Chen, and C. Chu, "Dcmlite Software," CSIRO Data Access Portal, PID, https://data.csiro.au/dap/landingpage?pid=csiro:9448.

22.   H. P. Wang, Y. S. Yang, Y. D. Wang, J. L. Yang, J. Jia, and Y. H. Nie, "Data-constrained modelling of an anthracite coal physical structure with multi-spectrum synchrotron X-ray CT," Fuel, vol. 106, pp. 219–225, 2013.

23.   T. Gureyv, Y. Nesterets, D. Thompson, S. Wilkins, A. Stevenson, and A. Sakellariou, "Toolbox for advanced X-ray image processing," in Advances in Computational Methods for X-Ray Optics II, C. Oleg, Ed., Proceedings of SPIE, pp. 8141–8110, 2011.

Chapter 3

Methane Steam Reforming on Supported Nickel Based Catalysts. Effect of Oxide Zro$_2$, La$_2$O$_3$ and Nickel Composition

Akila Belhadi[1], Mohamed Trari[2], Chérifa Rabia[1],
and Ouiza Cherifi[1,]

[1]Laboratory of Chemistry of Natural Gas, Faculty of Chemistry (USTHB), Algiers, Algeria

[2]Laboratory of Storage and Valorisation of Renewable Energies, Faculty of Chemistry (USTHB), Algiers, Algeria

[3]Université M'Hamed Bougara de Boumerdès, U.M.B.B., Boumerdès Department, Boumerdès, Algeria

ABSTRACT

The catalytic properties of Ni (4 and 10 wt%) supported on both La_2O_3 and ZrO_2 were investigated for the methane steam reforming reaction between 475°C and 700°C at atmospheric pressure. The catalysts were prepared by the impregnation method and characterized by several techniques (atomic absorption, BET method, X-ray diffraction and TG-TPO). The catalytic activity of Ni/support systems strongly depends on both of the nature and physico-chemical properties of the support. No deactivation was observed in catalytic systems, whatever the reaction temperature indicating high stability of the catalyst.

INTRODUCTION

The valorization of the natural gas via the conversion of methane presents attracting interest because of the abundance of natural gas and its low cost. The methane conversion to synthesis gas (mixture of CO and H_2) can be realized by different processes like the methane steam reforming (MSR) with H_2O [1-4], dry methane reforming with CO_2 [5-6] and methane oxidation with molecular oxygen [7-8]. The most catalysts are usually nickel-based systems because of their thermal stability and low cost [2, 9-11].

One of the major problems of these processes is the catalyst deactivation caused by carbon deposits formed on the surface which is related to high temperatures needed to activate the stable methane molecule. It has been reported in several works that the coke formation on Ni-based catalysts is sensitive to the acido basic character of support, metal-support interactions and metal crystalline structure. Supports with strong Lewis basicity as TiO_2, ZrO_2 [12-14] strong interactions between Ni and support lead to the formation of small Ni crystallites [15], and Ni in a spinel structure as $NiAl_2O_4$ [16] can minimize carbon deposit. Moreover, the Ni/Al_2O_3 systems with additives such as alkaline oxides (CaO, MgO) have also shown to be more resistant to coke formation [17-18].

However, some studies have shown that the deactivation of the catalyst is quite sensitive to the nature of carbon formed on the catalyst surface. Thus, the comparative study on two Ni/ -Al$_2$O$_3$ and Ni/SiC systems showed that the carbon as nanofilament formed with Ni/ -Al$_2$O$_3$, is originally the catalyst deactivation by blocking the active sites while the carbon in the nucleation form and growth observed on Ni/SiC favors the conversion of methane. The formation of different structures of carbon from the nickel surface was attributed to the existence of various metal-support interactions which modified the exposed faces of the metal [19].

In MSR reaction over supported Pt based systems, it was observed an enhancement of the catalytic activity that has been attributed to the high amount of carbon deposits around or near the Ni metal particles [20]. It has been underlined that the decomposition of CH$_4$ may take place on the metal particle, resulting in the formation of carbon and hydrogen and that carbon formed can partially reduce the support near the metal particles. Thus, the increase of carbon deposits around or near the particle metal should favor the methane conversion and the active sites are not encapsulated by carbon deposited. The aim of the present work is to report the catalytic behaviour of Ni/ZrO$_2$ and Ni/La$_2$O$_3$ with strong Lewis basicity of support and different loading of Ni (4 and 10 wt%), in the MSR reaction in the temperature range 475°C - 700°C. The catalysts have been characterized by BET, atomic absorption, X-ray diffraction (XRD) and TG-TPO techniques.

EXPERIMENTAL

Sample Preparation

ZrO$_2$ was prepared as described by Boulayt et al. [21] from zirconium n-propylate in propane-2-ol (70%). The ZrO$_2$-based precipitate was filtered off, washed several times, dried overnight at 110°C and then calcined at 300 °C/2h and 500°C/2h with a rate of 2°C/min under oxygen flow (1.2 L/h). The supported systems, Ni/MO [MO= La_2O_3 (Merck) and ZrO_2] were prepared by impregnation of the support MO with $Ni(NO_3)_2 \cdot 6H_2O$ aqueous solution (1 M). The mixture was stirred during 2 h at 80°C. The samples were dried at 80°C/12h, and calcined under air flow (1.2 L/h) at 300°C/2h and 500°C/2h with a rate of 4°C/min. The catalysts were then sieved to have a particle diameter less than 0.16 mm.

Characterization

Solid composition was determined by atomic absorption with a spectrometer type Perkin-Elmer 1100B. The specific surface area was determined by the BET method using nitrogen gas as absorbate on a surface analyzer (Coultronics 2100E) after pre-treatment of samples under vacuum at 200°C for 1 h (5°C/min) to have a clean surface. X-ray diffraction powder patterns were recorded with a $\theta/2\theta$ diffractometer (CGF) using Mo Kα radiation (λ = 0.70930 Å). The apparent sizes of nickel oxide and metal nickel particles were calculated from the Scherrer formula, L = 0.9λ/βcosθ, where β is the width of the most intense peak at half-height and θ the corresponding diffraction angle [22].

Activity Measurements

The MSR reaction carried out in a quartz fixed bed tubular reactor (L = 65 cm, f = 1 cm) under atmospheric pressure in the temperature range (450°C - 700°C), with on-line TCD chromatograph analysis (Hewlett-Packard 5730) using carbosieve B column, 100 - 200 mesh, of 2 m length and hydrogen as vector gas. A thermocouple was installed within the reactor, in contact with catalyst bed. Calcined Ni/MO sample (100 mg) was activated in situ overnight at 500°C under hydrogen flow (1.2 L/h). The gas feed consisted of methane and water in a ratio H_2O/CH_4 = 3.3. CH_4(10%)/Ar was introduced to the reaction zone by flowing through a water saturator maintained at 65°C with a flow rate of 1.2 L/h. Before each analysis, the reactants and products pass through a water-trap at $0°C$ to remove water. The conversion of CH_4 and product selectivities is calculated using following formulas:

$$\text{Conversion}_{CH_4}\,(\%) = \frac{n_{CH_4}^{in} - n_{CH_4}^{out}}{n_{CH_4}^{in}} \times 100$$

$$\text{Selectivity}_{CO}\,(\%) = \frac{n_{CO}^{out}}{n_{CH_4}^{in} - n_{CH4}^{out}} \times 100$$

$$\text{Selectivity}_{CO_2}\,(\%) = \frac{n_{CO_2}^{out}}{n_{CH_4}^{in} - n_{CH_4}^{out}} \times 100$$

$$\%C = \left(n_{CH_4}^{in} - \left(n_{CH_4}^{out} + n_{CO}^{out} + n_{CO_2}^{out} \right) \right) \times 100$$

$n°$: number of moles.

Coke Oxidation

In order to determine the carbon amount, the oxidation of coked catalyst was performed using MTB 10 - 8 Setaram Microbalance with relative and absolute sensitivities of 4.10 - 8 and 4.10 - 7 g respectively. The microbalance is linked to a computer via a Cobra interface. 30 mg of coked sample was pretreated at $50°C$ under vacuum (10 - 3 Pa) until stabilization of the weight. Molecular oxygen was introduced at a pressure of 200 mbar. The oxidation temperature rose to $550°C$ ($5°C/min$) and the sample was maintained at this temperature during 6 h.

RESULTS AND DISCUSSION

Catalytic Systems Characterization

The physical characteristics of the solids are summarised in Table 1 and XRD patterns of 4% Ni/La_2O_3 in Figure 1. The atomic absorption analysis shows that the composition of different systems is very close to the theoretical one.

The crystallite size was calculated from X-ray line broadening of NiO and that of Ni peaks (2 = 19.5° and 20° respectively) using the Scherrer equation. The results show that the support influence significantly the average size of Ni particles with ca. 12 nm and ca. 22 nm for Ni/La_2O_3 and Ni/ZrO_2 respectively, while for the NiO particles, the value is ca. 36 nm for both carriers. The formation of small Ni particles observed in presence of La_2O_3 can be associated to the stronger interactions between Ni, NiO and La_2O_3, more basic than ZrO_2, as observed in the case of Ni/MgO catalyst [23]. It is noted that the particle size of NiO and Ni appear to be independent of the deposited amount of active phase.

The specific surface areas of 4wt% and 10wt% Ni/ZrO_2 are similar with 86 and 88 m^2/g, slightly lower than that of the ZrO_2 support (93 m^2/g). La_2O_3 support has a very low surface area (1 m^2/g) that increases to 8 and 15 m^2/g after impregnation of 10 and 4 wt% Ni respectively. This increase may be due to the presence of both $La(OH)_3$ and $Ni(OH)_2$ phases examined by XRD analysis. Contrarily to Ni/ZrO_2 system, in presence of La_2O_3, the specific surface area decreases from 15 to 8 m^2/g with increasing of Ni percentage from 4 wt% to 10 wt%. This decrease could be explained by the formation of agglomerates on the support surface.

Table 1: Characteristics of Ni/support systems

Samples	S_{bet}(m²/g)	Ni Exp(wt%)	Ni theo(wt%)	XRD Result			
				D^3(nm)Ni o	D^3(nm)Ni	Before reaction	After reaction
La_2O_3	1						
$4Ni/La_2O_3$	15	4	4	36.4	11.7	La_2O_3, $La(OH)_3$	
$10Ni/La_2O_3$	8	10.3	10	35.2	11.7	NiO, $Ni(OH)_2$, $La(OH)_3$	Ni°, $Ni(OH)2 \cdot 0.75H2O$, $La(OH)_3$, C
						NiO, $La(OH)_3$	Ni, $La(OH)_3$, C
ZrO_2	93					ZrO_2	
$4Ni/ZrO_2$	86	4.3	4	35.8	21.5	NiO, ZrO_2	Ni, ZrO_2
$10Ni/ZrO_2$	88	9.98	10	35.3	21.6	NiO, ZrO_2	Ni°, ZrO_2

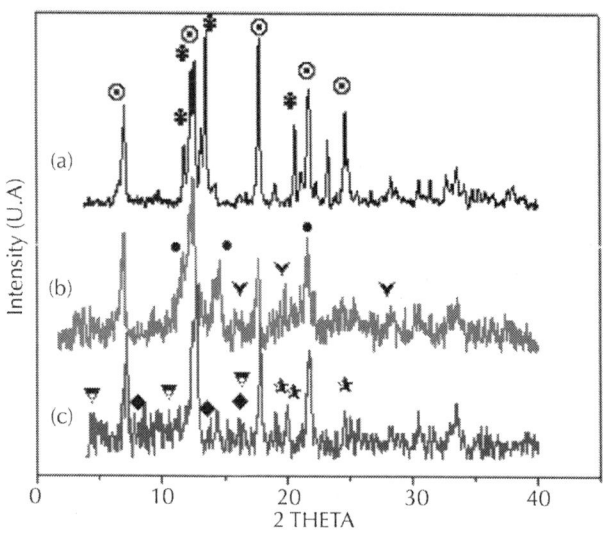

Figure 1: XRD patterns of support (a) La_2O_3 and catalysts 4(%) Ni/La_2O_3(b) before and (c) after reaction ⊚: $La(OH)_3$; *: La_2O_3; ⌄: Nio; ●: $Ni(OH)_2$;▼: $Ni(OH)_2$ 0.75H_2O; ✳: Ni°, ◆Camorphe.

After calcination at 500°C, the XRD pattern of 4 wt% Ni/La_2O_3 (Figure 1) shows peaks assigned to $Ni(OH)_2$, $La(OH)_3$ and NiO and no peak corresponding to La_2O_3 is observed. It is well known that La_2O_3 is highly hygroscopic at room temperature.

After MSR reaction, the presence of Ni° metallic species and that of support are visible in the patterns of Ni/ZrO_2 (Table 1), whereas in the case of Ni/ La_2O_3, in addition to the presence of Ni° metallic species, there appear peaks attributed to carbon, $Ni(OH)_2$ and $La(OH)_3$. The absence of peaks corresponding to carbon in presence of Ni/ZrO_2 catalyst could be due to an amorphous form of carbon.

Methane Steam Reforming Reaction

The catalytic performances of supported Ni systems in the MSR reaction were examined in the temperature range (475°C - 700°C), after reduction pretreatment under hydrogen flow at 500°C (1.2

L/h) for 12 h [24]. The MSR reaction over the catalysts leads to the formation of CO, CO_2, H_2 and carbon and the results are reported in Figures 2, 3 and Tables 2, 3.

Effect of Calcination Temperature

For the preparation of $4\%Ni/ZrO_2$ sample, two calcinations temperatures (500°C and 700°C) [25] were used to examine their effect on the catalytic performance. Figure 2 shows methane conversion and CO selectivity as a function of reaction temperature. After reduction pretreatment (H2/500°C/overnight), $4\%Ni/ZrO_2$ system leads to similar evolution of the methane conversion with reaction temperature for calcinations temperatures 500°C and 700°C. When the catalyst is calcined at 500°C, the CO formation is favoured at low reaction temperatures (<650°C) and from 650°C; the CO selectivity is the same for both calcinations temperatures. This result shows that the used calcination temperature, during the catalyst preparation, does not have a significant effect on the conversion whereas CO formation is favored when the calcination temperature is 500°C. So, the calcination temperature was fixed at 500°C for all studied systems.

Steady-State Performance

The catalytic activity of 4%wt Ni/support in the MSR reaction was examined in the temperature range (475°C - 700°C) after in situ pretreatment of the catalyst under hydrogen flow at 500°C overnight (Figure 3).

Similar evolutions of the methane conversion as a function of time-on-stream were obtained at different temperatures for 4%wt Ni/ZrO_2 catalyst.

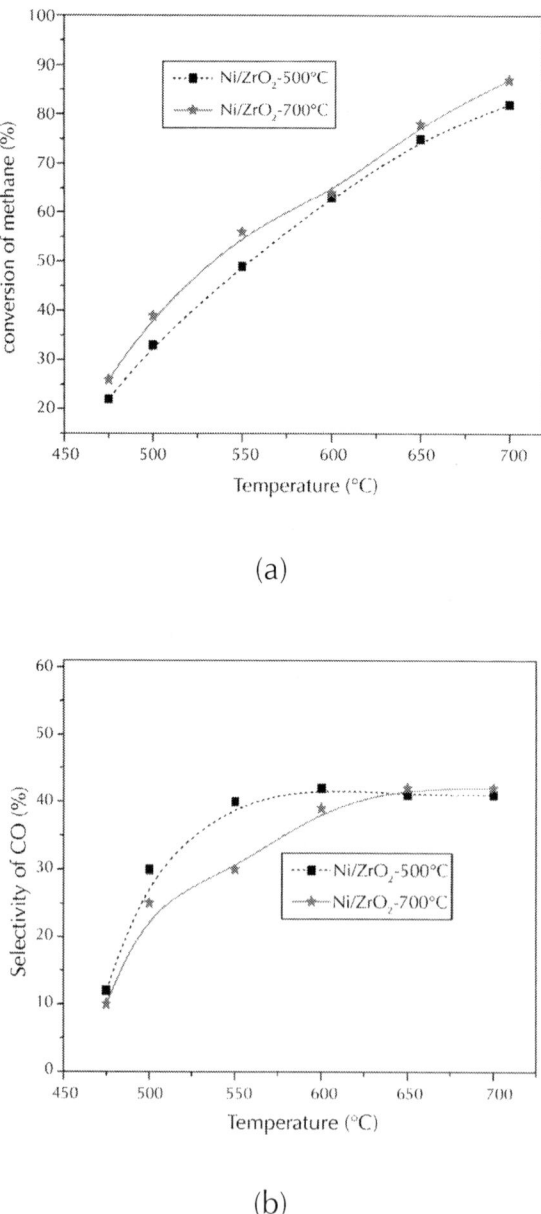

(a)

(b)

Figure 2: Methane conversion and CO selectivity for 4%Ni/ ZrO$_2$ calcined at 500°C and at 700°C at different reaction temperatures, m = 0.1 g, Tred = 500°C/H$_2$/over-night, d = 1.2 L·h^{-1}, H$_2$O/CH$_4$ = 3.3.

ZrO_2 with conversions of 22% - 82% against 22% - 72% and more selective toward CO with 30% - 46% against 29% - 40%. It is also noted that 10%wt Ni/ZrO_2 shows no activity at 475°C contrary to 4%wt Ni/ZrO_2 that displays a conversion of 22%. On the other hand, over 4%wt Ni/ZrO_2 catalyst, the amount of hydrogen did not change markedly with temperature (17.0 - 17.5 mmol/g·h), while on 10%wt Ni/ZrO_2, the hydrogen quantity increases substantially from 25.0 to 76.0 mmol/g·h.

Table 2: Catalytic activities of Ni/ZrO_2 catalyst

Catalysts	Temperature (°C)	Methane Conversion (%)	Selectivity Co (%)	Selectivity Co_2 (%)	%C	$H_2 10^{-3}$ (m01/gh)
4%ni/Zro₂	475	22	12	42	46	
	500	33	30	23	47	17.0
	550	49	46	6	48	17.2
	600	62	42	Traces	58	17.4
	650	75	41	0	59	17.5
	700	82	41	0	59	17.5
4%ni/Zro₂	475	Traces	-	Traces	-	
	500	22	29	30	41	25.0
	550	33	35	6	59	39.6
	600	48	36	3	61	61.4
	650	67	38	0Traces	62	67.9
	700	72	40	0	60	76.0

Table 3: Catalytic activities of Ni/La$_2$O$_3$ catalyst

Catalysts	Temperature (°C)	Methane Conversion (%)	Selectivity Co (%)	Selectivity Co$_2$ (%)	%C	H$_2$10^{-3}(m01/gh)
4%ni/Zro$_2$	475	-	-	-	-	-
	500	Traces	-	-	-	-
	550	20	50	35	15	26.1
	600	50	52	6	42	56.6
	650	79	35	2	63	67.4
	700	88	51	Traces	69	80.2
4%ni/Zro$_2$	475					
	500	Traces	Traces	-	-	-
	550	26	20	39	-	-
	600	51	34	26	22	36.2
	650	79	50	2	63	38.2
	700	88	52	traces	68	40.2

CO$_2$ is reaction product at low temperature with 42% of selectivity over 4 wt% Ni/ZrO$_2$ at 475°C and 30% in the case of 10 wt% Ni/ZrO$_2$ at 500°C. These values decrease with increasing temperature until they reach zero value above 600°C. On the contrary, the carbon is the major product at 550°C with selectivity varying between 48% - 59% for 4%Ni/ZrO$_2$ and from 500°C with selectivity varying between 41% - 62% for 10%Ni/ZrO$_2$. These results show that the increase of carbon deposits favored the methane conversion. Moreover, the selectivities toward CO and H$_2$ have practically not changed during reaction beyond 550°C. The comparison of the results of tables 2 and 3 shows that the nature of support has an effect on catalytic performance of Ni based catalyst, particularly on the product distribution. In contrast to Ni/ZrO_2 system, the amount of Ni supported on La_2O_3 has no effect on the activity of solid. Similar evolution conversions from 20% - 26% to 88% with increasing the reaction temperature from 550°C to 700°C, were observed for 4 wt% Ni/La_2O_3 and 10 wt% Ni/La_2O_3 while the effect of Ni content on the product distribution is more pronounced. Thus on 4 wt% Ni/La_2O_3, high CO selectivities were obtained at low reaction temperature (550°C and 600°C), 50 and 52% against 20% and 34% respectively for 10 wt% Ni/La_2O_3. The formation of hydrogen favored on 4 wt% Ni/La_2O_3 with a amount varying between 26.1 - 80.2 against 0.0 - 40. 4 mmol/g·h for 10 wt% Ni/La_2O_3. It is noteworthy that 4 wt% Ni/La_2O_3 is more selective than Ni/ZrO_2 catalyst.

The carbon is the major product beyond 650°C with selectivities of 63% - 69% for Ni/La_2O_3 catalysts while the CO_2 product is observed only in trace amounts (2% of selectivity) in the presence of 10 wt% Ni/La_2O_3 at 550°C. These results show also that the increase of carbon deposits favored the methane conversion as in the case of Ni/ZrO_2.

The enhancement of the activity (tables 2 and 3) and the stability (Figure 2) of both Ni/ZrO_2 and Ni/La_2O_3 on the MSR reaction can be attributed to the high amount carbon deposits around or near the Ni metal particles. These results are in agreement with those obtained by other authors on Pt/Al_2O_3, Pt/ZrO_2 and $Pt/Ce-ZrO_2$ systems [7-8,27-28]. It has been reported in these works that the decomposition of CH_4 takes place on the metal particle, resulting in the formation of carbon and hydrogen and that carbon formed can partially reduce the support near the metal particles. Thus, the increase of carbon deposits near the metal particles favors the methane conversion and the active sites (Ni metal) were not encapsulated by carbon deposited.

It was underlined that the catalyst deactivation is also sensitive to the nature of the support. Thus, the comparative study performed on two $Ni/ -Al_2O_3$ and Ni/SiC systems showed that the carbon as nanofilament was originally the catalyst deactivation by blocking the active sites while the carbon in the nucleation and growth form favored the conversion of methane. The formation of different structures of carbon from the nickel surface was attributed to the existence of different metal-support interactions which modified the exposed faces of the metal [19].

Determination of Coke Deposited

Temperature-programmed oxidation (TPO) coupled tothermogravimetric analysis (TG) was carried out on the Ni/support catalysts after 7 h of SRM reaction at 700°C (Figure 4). The TG curves of 4% Ni/ZrO_2 and 4% Ni/La_2O_3 (figure not shown) catalysts are similar with weight losses divided into two major events between 340°C and 500°C and between 350°C and 550°C respecttively attributed to CO_2 release, coming probably from two different forms of carbon. It has been reported that graphitic carbon was ignited at a higher temperature of around 500°C and reactive carbonaceous deposit or adsorbed carbon monoxide on the surface was ignited at a lower temperature below 400°C [29]. The results also revealed the formation of a higher amount of carbon on 4% Ni/ZrO_2 catalyst compared to the 4% Ni/La_2O_3 catalyst (35 wt% of CO_2 against 10 wt%).

TPO-TG analysis shows a different behavior between 10 wt% Ni/support and 4 wt% Ni/support. Thus, with Ni content of 10 wt%, gradual weight losses of ca. 20 wt% from 350°C to 550°C for Ni/ZrO_2 and of 65 wt% from 360°C to 500°C for Ni/La_2O_3 were observed.

These results indicate that the carbon formation depends on Ni loading and support nature. La_2O_3support favored the carbon deposited when Ni content is high contrarily to ZrO_2 support.

CONCLUSIONS

The obtained results showed that supported Ni (4 and 10 wt%) on ZrO_2 and La_2O_3 support exhibited high catalytic activity (72% - 88% of methane conversion) at 700°C and high stability for the steam reforming methane reaction to synthesis gas. The large amount of carbon depos- ited on the catalyst surface does not affect the activity of the Ni/ZrO_2 and Ni/ La_2O_3 systems, probably due to a form of carbon that is not detrimental to catalyst activity. On the other hand, the catalysts seem to develop a self-stabilization process during the reaction.

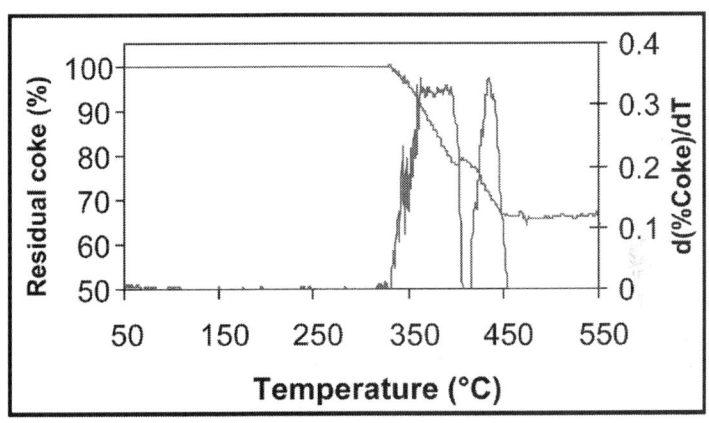

(a)

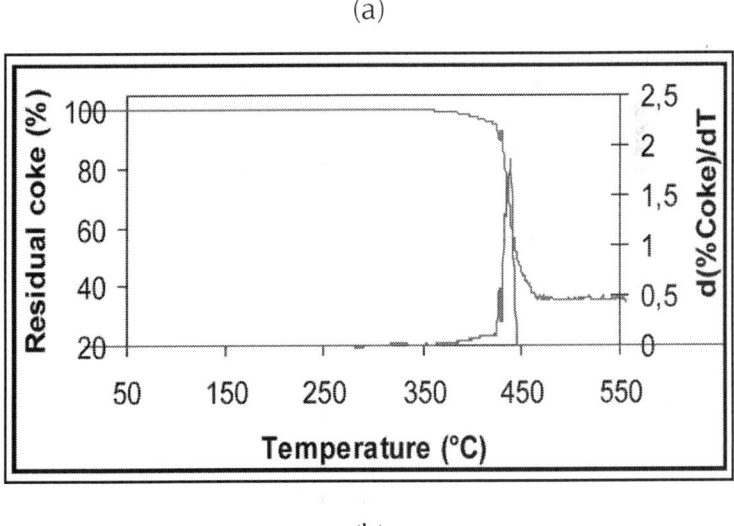

(b)

Figure 4: Residual coke (%) and the DTG during the TPO of: (a)—4% Ni/ ZrO_2 , (b)—10% Ni/La_2O_3.

ACKNOWLEDGEMENTS

We are thankful to Pr. Y. Boucheffa (UER de Chimie Appliquée, EMP) for his contribution for thermal analysis (TG, TPO).

REFERENCES

1.   B. Neumann and K. Jacob, "Equilibrium in Formation of Methane from Carbon Monoxide and Hydrogen, or from Carbon Dioxide and Hydrogen," Schrift für Elektrochemie, Vol. 30, 1924, pp. 557-576.

2.   J. R. Rostrup-Nielsen, "Catalytic Steam Reforming," In: J. R. Anderson and M. Boudart, Eds., Catalysis: Science and Technology, Springer-Verlag, New York, 1984, pp. 1-117. doi:10.1007/978-3-642-93247-2_1

3.   K. Hou and R. Hughes, "The Kinetics of Methane Steam Reforming over a Ni/a-Al$_2$O Catalyst," Chemical Engineering Journal, Vol. 82, No. 1-3, 2001, pp. 311-328.doi:10.1016/S1385-8947(00)00367-3

4.   S. Zhang, J. Wang and X. Wang, "Effect of Calcination Temperature on Structure and Performance of Ni/TiO$_2$- SiO$_2$ Catalyst for CO$_2$ Reforming of Methane," Journal of Natural Gas Chemistry, Vol. 17, No. 179, 2008, pp. 179- 183. doi:10.1016/S1003-9953(08)60048-1

5.   V. R Choudhary, B. S Uphade and A. S. Mamman, "Simultaneous Steam and CO$_2$Reforming of Methane to Syngas over NiO/MgO/SA-5205 in Presence and Absence of Oxygen," Applied Catalysis A: General, Vol. 168, No. 1, 1998, pp. 33-46.doi:10.1016/S0926-860X(97)00331-1

6.   J. H. Kim, D. J. Suh, T. J. Park and K. L. Kim, "Effect of Metal Particle Size on Coking during CO$_2$ Reforming of CH$_4$ over Ni-Alumina Aerogel Catalysts," Applied Catalysis A: General, Vol. 197, No. 2, 2000, pp. 191-200.

7. L. V. Mattos, E. R. de Oliveira, P. D. Resende, F. B. Noronha and F. B. Passos, "Partial Oxidation of Methane on Pt/Ce-ZrO$_2$ Catalysts," Catalysis Today, Vol. 77, No. 3, 2002, pp. 245-256. doi:10.1016/S0920-5861(02)00250-X

8. F. B. Noronha, E. C. Fendley, R. R. Soares, W. E. Alvarez and D. E. Resasco, "Correlation between Catalytic Activity and Support Reducibility in the CO$_2$ Reforming of Methane over Pt/Ce$_x$Zr$_{1-x}$O$_2$ Catalysts," Chemical Engineering Journal, Vol. 82, No. 1-3, 2001, pp. 21-31. doi:10.1016/S1385-8947(00)00368-5

9. H. S. Roh, K. W. Jun, W. S. Dong, J. S. Chang, S. E. Park and J. Yung-II, "Highly Active and Stable Ni/Ce-ZrO$_2$ Catalyst for H$_2$ Production from Methane," Journal of Molecular Catalysis A: Chemical, Vol. 181, No. 1-2, 2002, pp. 137-142. doi:10.1016/S1381-1169(01)00358-2

10. Y. Wang, Y. H. Chin, R. T. Rozmiarek, B. R. Johnson, Y. Gao, J. Watson, A. Y. L. Tonkovich and D. P. V. Wiel. "Highly Active and Stable Rh/MgO-Al$_2$O$_3$ Catalysts for Methane Steam Reforming," Catalysis Today, Vol. 98, No. 4, 2004, pp. 575-581.doi:10.1016/j.cattod.2004.09.011

11. T. Borowiecki, W. Gac and A. Denis, "Effects of Small MoO$_3$ Additions on the Properties of Nickel Catalysts for the Steam Reforming of Hydrocarbons: III. Reduction of Ni-Mo/Al$_2$O$_3$Catalysts," Applied Catalysis A: General, Vol. 270, No. 1-2, 2004, pp. 27-36.doi:10.1016/j.apcata.2004.03.044

12. T. Wu, Q. Yan and H. Wan, "Partial Oxidation of Methane to Hydrogen and Carbon Monoxide over a Ni/TiO$_2$ Catalyst," Journal of Molecular Catalysis A: Chemical, Vol. 226, No. 1, 2005, pp. 41-48. doi:10.1016/j.molcata.2004.09.016

13. V. R. Choudhary, S. Banerjee and A. M. Rajput, "Hydrogen from Step-Wise Steam Reforming of Methane over Ni/ZrO$_2$: Factors Affecting Catalytic Methane Decomposition and Gasification by Steam of Carbon Formed on the Catalyst," Applied Catalysis A: General, Vol. 234, No. 1-2, 2002, pp. 259-270. doi:10.1016/S0926-860X(02)00232-6

14. R. Takahashi, S. Sato, T. Sodesawa, M. Yoshida and S. Tomiyama, "Addition of Zirconia in Ni/SiO_2 Catalyst for Improvement of Steam Resistance," Applied Catalysis A: General, Vol. 273, No. 1-2, 2004, pp. 211-215. doi:10.1016/j. apcata.2004.06.033

15. Z. W. Liu, K. W. Jun, H. S. Roh, S. C. Baek, S. E. Park, "Pulse Study on the Partial Oxidation of Methane over $Ni/q\text{-}Al_2O_3$ Catalyst," Journal of Molecular Catalysis A: Chemical, Vol. 189, No. 2, 2002, pp. 283-293. doi:10.1016/S1381-1169(02)00365-5

16. N. Sahli, C. Petit, A. C. Roger, A. Kiennemann, S. Libs and M. M. Bettahar; "Ni Catalysts from $NiAl_2O_4$ Spinel for CO_2 Reforming of Methane," Catalysis Today, Vol. 113, No. 3-4, 2006, pp. 187-193. doi:10.1016/j.cattod.2005.11.065

17. S. AL-Ubaid, "The Activity and Stability of Nickel/Silica Catalysts in Water and Methane Reaction," Industrial & Engineering Chemistry Research, Vol. 27, No. 5, 1988, pp. 790-795. doi:10.1021/ie00077a013

18. M. V. Twigg, "Catalyst Handbook Mansson," 2nd Edition, Manson Publishing, London, 1994.

19. P. Leroi, B. Madani, C. Pham-Huu, M. J. Ledoux, S. Savin-Poncet and J. L. Bousquet, "Ni/SiC: A Stable and Active Catalyst for Catalytic Partial Oxidation of Methane," Catalysis Today, Vol. 91-92, 2004, pp. 53-58. doi:10.1016/j.cattod.2004.03.009

20. J. A. C. Ruiz, F. B. Passos, J. M. C. Bueno, E. F. SouzaAguiar, L. V. Mattos and F. B. Noronha, "Syngas Production by Autothermal Reforming of Methane on Supported Platinum Catalysts," Applied Catalysis A: General, Vol. 334, No. 1-2, 2008, pp. 259-267.doi:10.1016/j.apcata.2007.10.011

21. C. Lahousse, A. Aboulayt, F. Maugé, J. Bachelier and J. C. Lavalley, "Acidic and Basic Properties of ZirconiaAlumina and Zirconia-Titania Mixed Oxides," Journal of Molecular Catalysis,Vol.84,No.3,1993,pp.283-297.doi:10.1016/0304-5102(93)85061-W

22. C. R. Jung, J .Han, S. W. Nam, T. H. Lim, S. A. Hong and H. I. Lee, "Selective Oxidation of CO over CuOCeO$_2$ Catalyst: Effect of Calcination Temperature," Catalysis Today, Vol. 93-95, 2004, pp. 183-190. doi:10.1016/j.cattod.2004.06.039

23. Y. H. Wang and B. Q. Xu, "Comparative Study of Atmospheric and High Pressure CO$_2$Reforming of Methane over Ni/MgO-AN Catalyst," Catalysis Letters, Vol. 99, No. 1-2, 2005, pp. 89-96. doi:10.1007/s10562-004-0784-2

24. Belhadi and O. Cherifi, "Effet des Ajouts métalliques sur les Catalyseurs à Base de Nickel Supportés sur Silice, Dans la réaction de Vaporeformage du méthane," Journal de la Société Algérienne de Chimie, Vol. 19, No. 1, 2009, pp. 49-61.

25. F. Fally, V. Perrichon, H. Vidal, J. Kaspar, G. Blanco, J. M. Pintado, S. Bernal, G. Colon, M. Daturi and J. C. Lavalley, "Modification of the Oxygen Storage Capacity of CeO$_2$-ZrO$_2$Mixed Oxides after Redox Cycling Aging," Catalysis Today, Vol. 59, No. 3-4, 2000, pp. 373- 386. doi:10.1016/S0920-5861(00)00302-3

26. H. Vidal, J. Kaspar, M. Pijolat, G. Colon, S. Bernal, A. Cordón, V. Perrichon and F. Fally, "Redox Behavior of CeO$_2$-ZrO$_2$ Mixed Oxides: I. Influence of Redox Treatments on High Surface Area Catalysts," Applied Catalysis B: Environmental, Vol. 27, No. 1, 2000, pp. 49-63. doi:10.1016/S0926-3373(00)00138-7

27. S. M. Stagg-Williams and D. E. Resasco, "Effect of Promoters on Supported Pt Catalysts for CO$_2$ reforming of CH$_4$," Studies in Surface Science and Catalysis, Vol. 119, 1998, pp. 813-818. doi:10.1016/S0167-2991(98)80532-6

28. S. M. Stagg-Williams, F. B. Noronha, G. Fendley and D. E. Resasco, "CO$_2$ Reforming of CH$_4$ over Pt/ZrO$_2$ Catalysts Promoted with La and Ce Oxides," Journal of Catalysis, Vol. 194, No. 2, 2000, pp. 240-249. doi:10.1006/jcat.2000.2939

29. D. Li, T. Shishido, Y. Oumi, T. Sano and K. Takehira, "Self-Activation and Self-Regenerative Activity of Trace Rh-Doped Ni/Mg(Al)O Catalysts in Steam Reforming of Methane,"

Applied Catalysis A: General, Vol. 332, No. 1, 2007, pp. 98-109.doi:10.1016/j.apcata.2007.08.008.

Biogas Production from Various Typical Organic Wastes Generated in the Region of Cantabria (Spain): Methane Yields and Co-digestion Tests

Carlos Rico[1], Rubén Diego[2, 3], Agustín Valcarce[2], and José Luis Rico[4]

[1]Department of Water and Environment Science and Technologies, University of Cantabria, Santander, Spain

[2]Teican Medioambiental, Boo de Piélagos, Cantabria, Spain

[3]Department of Business Administration, University of Cantabria, Santander, Spain

[4]Department of Chemical and Process Engineering Resources, University of Cantabria, Santander, Spain

ABSTRACT

Batch trials were carried out to determine the methane potential yields of some typical organic wastes generated in the region of Cantabria (Spain): cocoa shell, cheese whey and sludges from dairy industry. Anaerobic co-digestion trials of these wastes with dairy manure were also investigated in batch at 35°C. Cheese whey obtained similar methane yields than dairy manure, between 17.5 and 19.3 L CH_4 kg^{-1} cheese whey compared with 18.0 L CH_4 kg^{-1} manure. Methane yields of various sludge samples collected from wastewater treatment facilities of dairy industries were influenced by its origin. Sludge samples from fat separation devices were the most productive in terms of specific methane yields compared with biological sludge from an aerobic reactor. Sludge samples from fat separator reached specific methane productivities of 350 and 388 L CH_4 kg^{-1} VS (10.5 and 24.1 L CH_4 kg^{-1} sludge), whereas biological sludge yielded 125 L CH_4 kg^{-1} VS (12.6 L CH_4kg^{-1} sludge). The methane potential of sludge samples was influenced by solids content. Cocoa shell resulted to be an interesting waste for anaerobic digestion due to its high VS content, yielding 195 L CH_4 kg^{-1} cocoa shell. It is a waste that can considerably improve methane yields in anaerobic co-digestion with dairy manure. However, at proportions of 10% cocoa shell, the process was hindered by hydrolysis of particulate matter. Anaerobic digestion at higher temperatures (thermophilic range) could be a better option for this kind of waste. Co-digestion of 5% cocoa shell with 35% dairy sludge and 60% dairy manure resulted in 80.5% higher methane production compared to anaerobic digestion of dairy manure alone.

INTRODUCTION

Cantabria is a small region in the Northern Coast of Spain that has a bovine population of around 280,000 livestock units (mainly milk) which generate about 4.5 million tons of semi-liquid manure (7% - 14% TS) annually. Much larger dairy farms have become more common since 1990, resulting in greater awareness and concern for the proper management of manure. Typical disposal methods for animal manure allow for the emission of methane, ammonia, particulate matter, unpleasant odors, volatile organic compounds and a variety of other air pollutants, which can damage the environment and pose risks to animal and human health [1] [2] . Moreover, when manure is not properly managed it can cause severe environmental problems such as eutrophication of water bodies due to its high organic matter, nitrogen and phosphorous concentrations [3] [4] . Anaerobic digestion of animal manure is a well-known technology [5] that allows converting these concerns in two valuable products: biogas, a renewable fuel and the digested manure, with improved fertilizer characteristics [6] . Refining digested manure into concentrated fertilizers can also stimulate the distribution of surplus nutrient to other zones that have nutrient deficit, reducing environmental impacts caused by uncontrolled manure land application. On the other hand, anaerobic digestion also fulfills the goal of reducing greenhouse gas (GHG) emissions [7] . During storage in manure pits, the submerged manure generates methane, a GHG with 21 times the global warming potential of carbon dioxide, according to the Intergovernmental Panel on Climate Change (IPCC). Methane emissions can be avoided by processing manure in biogas facilities where methane can be recovered and converted in green energy.

Application of anaerobic digestion can be extended for almost any kind of organic waste or biomass, such as the biodegradable part of municipal solid waste, sewage sludge, wastes and wastewaters from food industry, energy crops, etc. In the case of organic wastes and wastewaters, many of them are also sources of methane emissions if not managed properly [8]. However, anaerobic digestion of organic wastes can be hindered due to the presence of toxic compounds for anaerobic microorganism [9]. High organic content and low buffer capacity can also lead to inhibitory effects on anaerobic digestion due to acidification or formation of intermediate toxic compounds, as has been reported for slaughterhouse waste and cheese whey [10] [11] . In general, the characteristics of dairy manure and organic wastes are opposite for anaerobic digestion. Whereas methane yields for some organic wastes can reach several hundred cubic meters per ton of waste [12] [13] , the fermentation of manure alone results in lower methane yields due to the high water content and moderate anaerobic biodegradability of about 45% - 50% [14] . However, the high water content and buffer capacity of manure, as well as trace elements, have a positive effect on process stability. For this reason, most of the biogas plants are operated today by co-digestion (anaerobic digestion in which two or more substrates are mixed) of manure together with organic wastes [15]. Dilution of toxic compounds, increased load of biodegradable organic matter, improved balance of nutrients, synergistic effect of microorganisms and better biogas yields have been reported as potential benefits for the co-digestion process [16] . Therefore, anaerobic co-digestion has become a good strategy for both the waste treatment and the production of renewable fuels.

In the last decades, the anaerobic digestion process has widely developed all across the majority of the European countries, due to new trends to promoting the production of biogas from agro-industrial wastes. The EU- countries where the agricultural biogas plants are most developed are Germany, Denmark, Austria and Sweden [6]. However, in spite of environmental and socio-economic benefits of biogas plants, agro-industrial anaerobic digestion technology has not developed in Spain mainly due to the low price paid for electric energy produced by biogas plants (less than 0.15 per KWh-el). According to Alfonso et al. [17] , Cantabria has an available potential of 61 million m^3 per year of agro-industrial biogas. This volume of biogas would represent an electricity production of 139 GWh/year in combined heat and power (CHP) production systems, and the 5% of total electricity consumed in 2012 in Cantabria. According to the Spanish Statistical Office, the food industry sector represents 15.4% of total industry volume of Cantabria. It generates a variety of organic wastes and other by- products that could be used as biomass energy sources. In terms of amount, the most important are those derived from the manufacture of dairy products, canned fish and other products from cocoa, chocolate, coffee, bakery and pastry industry. At present, many of these wastes and by-products have no market and are not receiving bio- treatments for resources recovery. For instance, only 42.6% of 15,600 tons per year of cheese whey produced in the region were managed in 2005, but mainly as animal feed [18] . In 2011, 16,900 tons of cheese whey were produced in Cantabria with similar management strategies.

In the present work, the methane potential of various organic wastes produced in Cantabria has been determined. The aim of this study was not only to determine individual methane yields, but also to study the synergies of co-digestion with dairy manure.

METHODS

Substrates and Inoculum

Four typical organic wastes produced in Cantabria were tested to determine their methane potential yields: cocoa shell, cheese whey, sludge from dairy industry and dairy manure. Cocoa shell (CS), a solid waste, was proceeded from a dairy factory that manufactures dairy milk chocolate products. It was crushed to reduce particle size. Three samples of sludge from dairy industry were taken from two different factories. The sludge samples were collected from the wastewater treatment facilities. Two of the sludge samples (SL1 and SL2) were taken from the fat separator previous to biological treatment of two different factories. The other sludge sample was the dewatered biological sludge (DSL1) taken from the factory where SL1 was collected. The biological sludge proceeded from the wastewater aerobic reactor and was dewatered by centrifuging. Cheese whey (CW) samples originated from three different cheese production factories. Dairy manure (M) sample proceeded from an intensive dairy farm. Liquid and semi-liquid substrates (SL, DSL, CW and M) were stored at 4°C prior to use.

The anaerobic inoculum was collected from a pilot CSTR digester that processed the screened liquid fraction of dairy manure [19]. To ensure degradation of remaining degradable organic matter in inoculums, it was stored, allowing the release of gas at 35°C for a week. Characteristics of organic wastes and inoculum are shown in Table 1.

Experimental Design

Two types of test were performed, a Biochemical Methane Potential (BMP) and co-digestion experiment tests. The first was made to determine the ultimate methane yields of the different wastes alone. The second was made to assess different co-digestion ratios of manure with several mixtures of wastes. Both experiments were performed in batch.

All the tests were performed in triplicate using 500-mL serum bottles capped with rubber septum sleeve stoppers as reactors. Gas production was determined by pressure measurement. The pressure was taken from the headspace of the reactors through the septum with a syringe connected to a digital pressure sensor with silicon measuring cell (ifm, type PN78, up to 2 bar). Biogas samples were also taken through the septum by a needle connected to a syringe. All the reactors were manually stirred once a day. After set-up of the reactors, Nitrogen was flushed to remove the air in the headspace of the bottles. Thereafter, all the reactors were incubated at 35°C. Gas produced in each reactor was determined daily. Three blanks with water and inoculum were also tested to measure methane potential of inoculum. Results are expressed as means subtracting methane production from the blanks.

Biochemical Methane Potential (BMP) Tests

Each reactor was fed with 250 g of a mixture of inoculum and substrate with a $VS_{inoculum}/VS_{substrate}$ ratio of 2 to minimize diffusion limitation and to avoid acidification or toxicity inhibition. The headspace of each reactor was calculated by subtracting the volume of the mix inoculum-substrate from the volume of the reactor. As can be deducted from data in Table 1, the sample that required the highest amount of inoculum was CS. In this case, 246.7 g of inoculum and 3.3 g of CS were transferred to the reactors. On the contrary sample SL2 required 183.97 g of inoculum and 66.1 g of SL2.

Co-digestion Tests

The co-digestion tests were also carried out in 500-mL serum bottles. First, co-digestion of manure with only one type of substrate (bi-substrate) was performed with a low inoculum-substrate ratio. For cheese whey co-di- gestion tests, mixtures with 10%, 20% and 30% of cheese whey based on mass were tested.

Table 1: Characteristics of substrates and inoculum used in batch tests

Sample	TS (%)	VS (%)	COD (g L^{-1})	NH_4^+ -N (g kg^{-1})	NKT-N (g kg^{-1})	P$_T$ (g kg^{-1})	Alkalinity (g CaCO$_3$ L^{-1})
CS	95.5	87.4	-	-	3.8	3.7	-
SL1	8.0	6.2	-	-	4.8	1.0	-
SL2	4.0	3.2	-	-	2.3	0.5	-
DSL1	13.2	10.8	-	-	9.5	0.5	-
CW1	6.8	5.5	64.6	0.28	1.3	0.3	0.6
CW2	5.6	5.2	61.2	0.18	0.75	0.2	0
CW3	6.5	5.5	68.6	0.25	1.1	0.3	0
M	11.5	9.3	-	0.78	4.8	0.6	21.1
Inoculum	3.8	2.3	28.8	1.8	2.5	0.6	30.8

For dairy sludges, batch trials were carried out with 40% of sludge and 60% of manure. Finally, cocoa shell and manure co-diges- tion tests were assayed at 5% and 10% CS based on mass. For these tests, 200 g of substrate mixture and 50 g of inoculum were used. The main objective of this trial was to assay the stability and limitations of co-digestion process. Subsequently, the co-digestion tests were performed with different mixtures of manure and organic wastes (multi-substrate) with higher inoculums substrate ratios. The mixtures of wastes consisted of dairy sludges; dairy sludges and cheese whey; dairy sludges, cheese whey and cocoa shell, and finally a mixture with the wastes that yielded the highest amount of methane. For these tests, a $VS_{inoculum}/VS_{substrate}$ ratio of 1 was used.

Analytical Techniques

Biogas composition was measured on a 2 m Poropak T column in a HP 6890 GC System with helium as the carrier gas and a TCD detector. Methane volumes are expressed at standard conditions (STP: 0°C, 1 atm). All other analyses (TS, VS, COD, Total Kjeldahl Nitrogen (TKN-N), Ammonia Nitrogen (NH_4^+ -N), and Total Phosphorous (P_T) were performed according to Standard Methods [20] .

RESULTS AND DISCUSSION

Characteristics of Substrates

From data in Table 1, organic wastes can be classified in two types: Cocoa shell, which is a solid waste and the rest, which are liquid (cheese whey) or semi-liquid (manure and dairy sludges). The higher VS content of cocoa shell (87.4%), compared with the other substrates (3.2% - 10.8%) potentially represents much higher energy content. However due to low water content, anaerobic digestion of cocoa shell alone would not be feasible unless it is diluted with water or co-digested with other substrate, such as manure. Regarding the rest of parameters, Nitrogen content of DSL1 could lead to ammonia inhibition [21] . Alkalinity is also an important parameter for anaerobic digestion. For these tests, alkalinity content of inoculum and manure seems to be enough to ensure stability of the process.

BMP Test

Methane and specific methane yields of organic wastes assayed in this work are shown in Figure 1. Cocoa shell had the highest methane yield, 195 L CH_4 per kg. The other wastes yielded considerably lower amounts of methane due to their lower solids content. Cheese whey samples yielded similar methane production values: 18.5, 17.5 and 19.3 L CH_4 per kg cheese whey for samples CW1, CW2 and CW3 respectively. Sludges from dairy factories gave different values, the best sample, in terms of methane potential was the SL1 (24.1 L CH_4 kg^{-1}), that corresponds with the sludge obtained from the fat separator of a dairy factory. The other sludge collected from the fat separator in other dairy factory (SL2) yielded a lower amount (10.5 L CH_4 kg^{-1}), but the reason for this difference can be found in the VS content, 6.2% for SL1 and 3.2% for SL2. The biological dewatered sludge (DSL1) produced 12.6 L CH_4 kg^{-1}. The methane yield for manure was a typical value for this kind of substrate, 18.0 L CH_4 kg^{-1}.

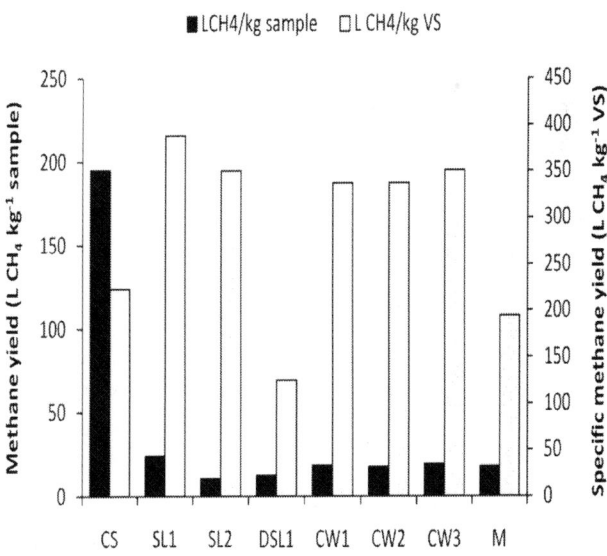

Figure 1: Methane and specific methane yields from organic wastes.

By analyzing the specific methane yields, it can be observed that SL1, SL2 and the three cheese whey samples obtained the highest values, between 337 and 388 L CH_4 kg^{-1} VS. That indicates the high biodegradability of these substrates. The three cheese whey samples produced similar values: 337, 337 and 351 L CH_4 kg^{-1} VS for CW1, CW2 and CW3 respectively. SL1 and SL2 yielded 388 and 350 L CH_4 kg^{-1} VS respectively. On the contrary, the dewatered biological sludge (DSL1) only yielded 125 L CH_4 kg^{-1} VS which is a low value compared with the other sludge samples. Its high Nitrogen content could be the reason for this low specific methane productivity. However, in BMP test the sample was highly diluted with the anaerobic inoculum to avoid toxicity effects which indicates that biodegradability of that sample was low. The reason for this low value can be that readily biodegradable organic matter was removed in the wastewater treatment plant, resulting in a dewatered sludge with a low specific methane yield. However, due to its high solids content (10.8% VS), methane production per kg of sludge reached a value (12.6 L CH_4 kg^{-1}) higher than that for SL2. The specific methane yield of cocoa shell (223 L CH_4 kg^{-1} VS) was low compared with SL1, SL2 and cheese whey samples, but higher than that of manure (194 L CH_4 kg^{-1} VS).

Co-digestion Tests

Bi-substrate Co-digestion Tests

Based on individual methane potentials, the estimated methane yields of mixtures were assessed and compared with the experimental values. The sample 3 of cheese whey (CW3) was selected as the one with the highest methane potential. In Figure 2, cumulative methane yields of cheese whey and manure co-digestion trials are shown. Mixtures with 10%, 20% and 30% cheese whey were tested. Since methane yields of cheese whey and dairy manure were similar, differences in methane yields of mixtures with 10%, 20% and 30% CW3 were small. Experimental values were close to theoretical values determined for individual substrates. Anaerobic co-diges- tion of cheese whey did not improve methane potential of manure, but co-digestion allowed digesting a difficult substrate such as the cheese whey without alkalinity addition.

Cumulative methane yields of cocoa shell and manure co-digestion trials are shown in Figure 3. For co-di- gestion of cocoa shell with manure, after 80 days both 5% and 10% samples did not reach their theoretical values based on individual trials. It can be observed that during the first 44 days, the methane production from sample with 5% CS was higher than that for 10% CS sample. In addition, cumulative methane production was still increasing after 80 days. The methane yield of 5% CS sample (25.1 L CH_4 kg^{-1} mixture) was close to the theoretical value (26.8 L CH_4 kg^{-1} mixture) but for the 10% CS sample the experimental value (30.5 L CH_4 kg^{-1} mixture) was further than the theoretical value (35.7 L CH_4 kg^{-1} mixture). The reason is that cocoa shell is a solid waste and anaerobic digestion is limited by hydrolysis of particulate matter. The higher cocoa shell ratio, the higher methane potential but hydrolysis step hinders the process at higher cocoa shell ratios. Moreover, dairy manure is a substrate with high suspended solids content and the amount of inoculum employed was also low (20% based on mass).

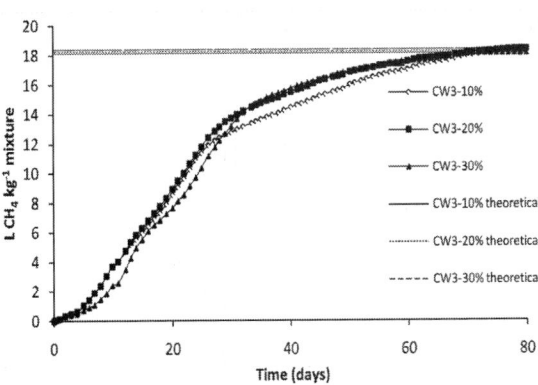

Figure 2: Cumulative methane production from co-digestion of mixtures of cheese whey and dairy manure.

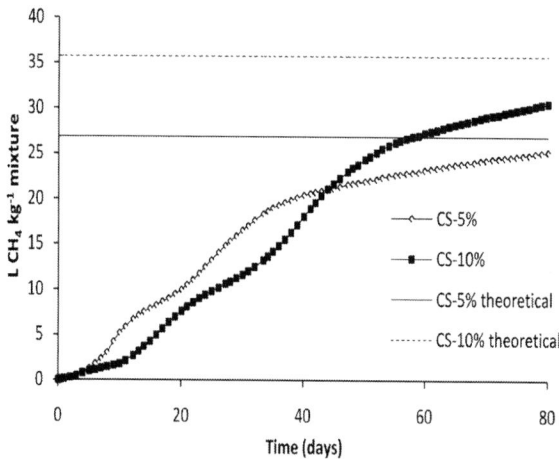

Figure 3: Cumulative methane production from co-digestion of mixtures of cocoa shell and dairy manure.

Thermophilic anaerobic co-digestion at higher temperature could be an alternative to enhance the hydrolysis step. The high methane potential of cocoa shell allowed increasing the methane potential of manure.

The effect of higher cocoa shell ratio in the co-digestion process is depicted in Figure 4, where specific methane yields of both mixtures are shown. Methane yields of 124 and 68 L CH_4 kg^{-1} VS were obtained after 30 days of digestion for CS-5% and CS-10% respectively. This difference was reduced with time, but shows difficulties in anaerobic co-digestion of cocoa shell at 10% ratio as low methane production rates suggest.

Anaerobic co-digestion of dairy sludge and manure was performed at 40% sludge ratios based on mass. As Figure 5 shows, despite the high ratios of sludge, all the samples reached their theoretical yields based on individual methane yields. Co-digestion samples SL2-40% and DSL1-40% resulted in lower methane yields than manure due to the lower methane potentials of SL2 and DSL1. Co-digestion of manure with SL1 allowed increasing methane yield of manure alone, 20.6 L CH_4 per kg mixture compared to 18.0 L CH_4 per kg manure.

Multi-substrate Co-digestion Tests

Proportions of wastes in multi-substrate co-digestion tests are specified in Table 2. All the samples were composed of 60% dairy manure and 40% of a mixture of different wastes, based on mass. Sample 1 (S1) was com- posed of 60% of dairy manure and a mix of the three dairy sludges.

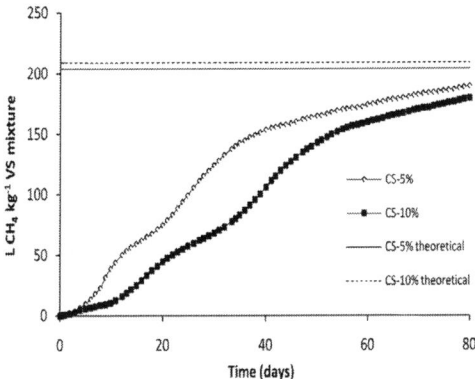

Figure 4: Cumulative specific methane production from co-digestion of mixtures of cocoa shell and dairy manure.

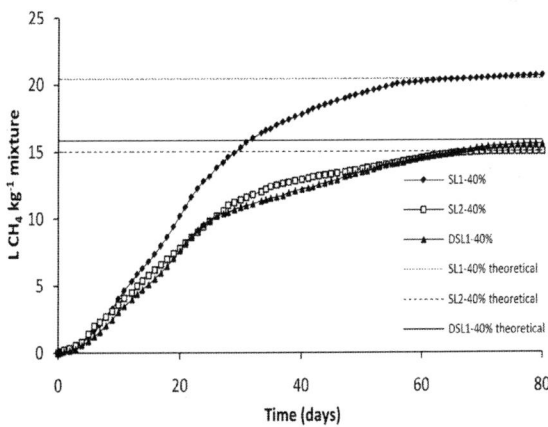

Figure 5: Cumulative methane production from co-digestion of mixtures of dairy sludge and dairy manure.

Table 2: Proportions of wastes used in multi-substrate mixtures

Sample	M (%)	SL1 (%)	SL2 (%)	DSL1 (%)	CW3 (%)	CS (%)
S1	60	13.33	13.33	13.33	-	-
S2	60	6.66	6.66	6.66	20	-
S3	60	6	6	6	20	2
S4	60	35	-	-	-	5

For sample 2 (S2), cheese whey was included as co-substrate in the mixture. In sample 3 (S3), cocoa shell was added at 2% mass ratio. Sample 4 (S4) was done with the aim to obtain a maximum methane yield. It was composed of 60% dairy manure, 35% SL1 and 5% cocoa shell.

The methane yields for the different multi-substrate mixtures are presented in Figure 6. The highest methane yield was obtained by sample 4 that composed of manure (60%), dairy sludge (SL1, 35%) and cocoa shell (5%) with a yield of 32.5 L CH_4 kg^{-1} mixture.

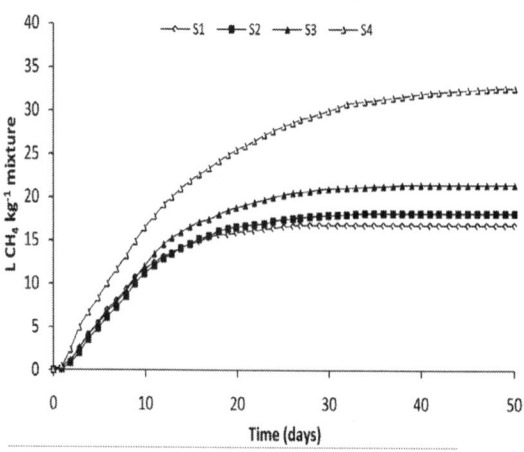

Figure 6: Cumulative methane production from multi-substrate co-digestion mixtures.

Samples 1 (S1) and 2 (S2) obtained similar methane yields, 16.7 and 18.1 L CH_4 kg^{-1} mixture, respectively. Higher yield for sample 2 was caused by incorporation of cheese whey into the mixture. Compared to samples 1 and 2, sample 3 reached a higher yield (21.4 L CH_4 kg^{-1} mixture) due to the presence of cocoa shell.

This trial provides good evidence that all the wastes studied are suitable for anaerobic co-digestion with dairy manure. In terms of methane potential, wastes from dairy industry (cheese whey and dairy sludges) did not improve methane yield from manure, except the sludge sample SL1. In this sense, sludges collected from fat separation facilities exhibited high specific methane yields, between 350 and 388 L CH_4 kg^{-1} VS. In spite of low methane yields, anaerobic co-digestion of organic wastes from dairy industry with dairy manure would be a good management strategy. Regarding cocoa shell, it is an interesting waste for anaerobic co-digestion that can considerably improve methane yield due to its high VS content. However, at proportions of 10% cocoa shell the process was hindered by hydrolysis of particulate matter. Anaerobic digestion at higher temperatures (thermophilic range) could be a better option for this kind of waste.

CONCLUSIONS

It has been shown that the wastes tested constitute a good substrate for anaerobic co-digestion with dairy manure. On the one hand, the management of these wastes would cease to be a problem. On the other hand, by means of anaerobic digestion their resources could be recovered to produce biogas, contributing to reducing the dependence from fossil fuels. Liquid and semi-liquid wastes from dairy industry showed methane potentials in the same range as dairy manure. On the contrary, cocoa shell has demonstrated to be a substrate with methane potential much higher than that of manure. In mesophilic conditions, anaerobic co-digestion with 5% cocoa shell and 35% dairy sludge based on mass allowed increasing 80.5% methane yields of manure alone.

ACKNOWLEDGMENTS

We acknowledge the technical staff of Teican Medioambiental for their contribution to this work. This research was supported by Iberdrola Renovables and Ocyener in the framework of the Concurso Eólico in the Autonomous Community of Cantabria.

REFERENCES

1. Lesschen, J.P., van den Berg, M., Westhoek, H.J., Witzke, H.P. and Oenema, O. (2011) Greenhouse Gas Emission Profiles of European Livestock Sectors. Animal Feed Science and Technology, 166, 16-28. http://dx.doi.org/10.1016/j.anifeedsci.2011.04.058

2. Vander Zaag, A.C., MacDonald, J.D., Evans, L., Vergé, X.P.C. and Desjardins, R.L. (2013) Towards an Inventory of Methane Emissions from Manure Management That Is Responsive to Changes on Canadian Farms. Environmental Research Letters, 8, 13 p.

3. Ball Coelho, B., Murray, R., Lapen, D., Topp, E., Bruin, A. and Khan, B. (2012) Phosphorus and Sediment Loading to Surface Waters from Liquid Swine Manure Application under Different Drainage and Tillage Practices. Agricultural Water Management, 104, 51-61. http://dx.doi.org/10.1016/j.agwat.2011.10.020

4. De Vries, J.W., Groenestein, C.M. and De Boer, I.J.M. (2012) Environmental Consequences of Processing Manure to Produce Mineral Fertilizer and Bio-Energy. Journal of Environmental Management, 102, 173-183.http://dx.doi.org/10.1016/j.jenvman.2012.02.032

5. Nasir, I.M., Mohd Ghazi, T.I. and Omar, R. (2012) Anaerobic Digestion Technology in Livestock Manure Treatment for Biogas Production: A Review. Engineering in Life, 12, 258-269. http://dx.doi.org/10.1002/elsc.201100150

6.   Holm-Nielsen, J.B., Al Seadi, T. and Oleskowicz-Popiel, P. (2009) The Future of Anaerobic Digestion and Biogas Utilization. Bioresource Technology, 100, 5478-5484.http://dx.doi.org/10.1016/j.biortech.2008.12.046

7.   Cuéllar, A.D. and Webber, M.E. (2008) Cow Power: The Energy and Emissions Benefits of Converting Manure to Biogas. Environmental Research Letters, 3, 8 p.

8.   Abbasi, T., Tauseef, S.M. and Abbasi, S.A. (2012) Anaerobic Digestion for Global Warming Control and Energy Generation. An Overview. Renewable and Sustainable Energy Reviews, 16, 3228-3242. http://dx.doi.org/10.1016/j.rser.2012.02.046

9.   Chen, Y., Cheng, J.J. and Creamer, K.S.(2008) Inhibition of Anaerobic Digestion Process: A Review. Bioresource Technology, 99, 4044-4064.http://dx.doi.org/10.1016/j.biortech.2007.01.057

10. Ergüder, T., Tezel, U., Güven, E. and Demirer, G.N. (2001) Anaerobic Biotransformation and Methane Generation Potential of Cheese Whey in Batch and UASB Reactors. Waste Management, 21, 643-650. http://dx.doi.org/10.1016/S0956-053X(00)00114-8

11. Palatsi, J., Viñas, M., Guivernau, M., Fernandez, B. and Flotats, X. (2011) Anaerobic Digestion of Slaughterhouse Waste: Main Process Limitations and Microbial Community Interactions. Bioresource Technology, 102, 2219-2227.http://dx.doi.org/10.1016/j.biortech.2010.09.121

12. Dinuccio, E., Balsari, P., Gioelli, F. and Menardo, S. (2010) Evaluation of the Biogas Productivity Potential of Some Italian Agro-Industrial Biomasses. Bioresource Technology, 101, 3780-3783. http://dx.doi.org/10.1016/j.biortech.2009.12.113

13. Hejnfelt, A. and Angelidaki, I. (2009) Anaerobic Digestion of Slaughterhouse By-Products. Biomass and Bioenergy, 33, 1046-1054.http://dx.doi.org/10.1016/j.biombioe.2009.03.004

14. Rico, J.L., García, H., Rico, C. and Tejero, I. (2007) Characterisation of Solid and Liquid Fractions of Dairy

Manure with Regard to Their Component Distribution and Methane Production. Bioresource Technology, 98, 971-979. http://dx.doi.org/10.1016/j.biortech.2006.04.032

15. Weiland, P. (2006) Biomass Digestion in Agriculture: A Successful Pathway for the Energy Production and Waste Treatment in Germany. Engineering in Life Sciences, 6, 302-309.http://dx.doi.org/10.1002/elsc.200620128

16. Khalid, A., Arshad, M., Anjum, M., Mahmood, T. and Dawson, L. (2011) The Anaerobic Digestion of Solid Organic Waste. Waste Management, 31, 1737-1744.http://dx.doi.org/10.1016/j.wasman.2011.03.021

17. Alfonso, D., Brines, N., Peñalvo, E., Vargas, C., Pérez Navarro, A., Gómez, P., Pascual, A. and Ruiz, B. (2010) Cuantificación de materias primas para producción de biogás (PSE-Probiogas). http://213.229.136.11/bases/ainia_probiogas.nsf/0/FEB62 601BC95C8D6C125773D00394446/$FILE/Resumen_inf_ cuantificacion.pdf .

18. Villar, A. (2005) Situación y perspectivas de la gestión de los sueros de quesería generados en Cantabria (Centro de Investigación y Formación Agraria de Cantabria).http://www. cifacantabria.com/Documentacioncifa/download.php?sess= 0&parent=18&expand=1&order=name&binary=1&id=13).

19. Rico, C., Rico, J.L., Tejero, I., Muñoz, N. and Gómez, B. (2011) Anaerobic Digestion of the Liquid Fraction of Dairy Manure in Pilot Plant for Biogas Production: Residual Methane Yield of Digestate. Waste Management, 31, 2167-2173.http://dx.doi.org/10.1016/j.wasman.2011.04.018

20. APHA (1998) Standard Methods for the Examination of Water and Wastewater. 18th Edition, American Public Health Association, Washington, DC.

21. Bhattacharya, S.K. and Parkin, G.F. (1989) The Effect of Ammonia on Methane Fermentation Processes. Journal— Water Pollution Control Federation, 61, 55-59.

Chapter 5

Relationships between Water and Gas Chemistry in Mature Coalbed Methane Reservoirs of the Black Warrior Basin

Jack C. Pashin[a], Marcella R. McIntyre-Redden[b], Steven D. Mann[b], David C. Kopaska-Merkel[b], Matthew Varonka[c], and William Orem[c]

[a]Boone Pickens School of Geology, Oklahoma State University, Stillwater, OK 74074, USA

[b]Geological Survey of Alabama, Tuscaloosa, AL 35486-6999, USA

[c]U.S. Geological Survey, 12201 Sunrise Valley Drive, Mail Stop 956, Reston, VA 20192-0002, USA

ABSTRACT

Water and gas chemistry in coalbed methane reservoirs of the Black Warrior Basin reflects a complex interplay among burial processes, basin hydrodynamics, thermogenesis, and late-stage microbial methanogenesis. These factors are all important considerations for developing production and water management strategies. Produced water ranges from nearly potable sodium-bicarbonate water to hypersaline sodium-chloride brine. The hydrodynamic framework of the basin is dominated by structurally controlled fresh-water plumes that formed by meteoric recharge along the southeastern margin of the basin. The produced water contains significant quantities of hydrocarbons and nitrogen compounds, and the produced gas appears to be of mixed thermogenic-biogenic origin.

Late-stage microbial methanogenesis began following unroofing of the basin, and stable isotopes in the produced gas and in mineral cements indicate that late-stage methanogenesis occurred along a CO_2-reduction metabolic pathway. Hydrocarbons, as well as small amounts of nitrate in the formation water, probably helped nourish the microbial consortia, which were apparently active in fresh to hypersaline water. The produced water contains NH_4^+ and NH_3, which correlate strongly with brine concentration and are interpreted to be derived from silicate minerals. Denitrification reactions may have generated some N_2, which is the only major impurity in the coalbed gas. Carbon dioxide is a minor component of the produced gas, but significant quantities are dissolved in the formation water. Degradation of organic compounds, augmented by deionization of NH_4^+, may have been the principal sources of hydrogen facilitating late-stage CO_2 reduction.

INTRODUCTION

Natural gas is produced from coal by lowering reservoir pressure, and this lowering is achieved primarily by reducing the hydrostatic pressure component through the coproduction of formation water (e.g., McKee et al., 1988 and Seidle, 2011). Accordingly, significant volumes of water are produced from many coalbed methane (CBM) reservoirs, and the quantity and composition of this water varies greatly (e.g., Ayers and Kaiser, 1994, Pashin et al., 1991 and Reddy, 2010). Management of coproduced water is thus a central issue in CBM development, because the water must be processed and disposed in an environmentally responsible manner. The quantity and quality of formation water in coal, moreover, is intimately related to the architectural framework and geologic history of the host sedimentary basin. For this reason, basin hydrodynamics is thought to have a significant influence on the generation and alteration of natural gases in coal. Thus, basin hydrology and geochemistry are fundamental considerations in the design and implementation of reservoir development protocols (e.g., Kaiser et al., 1994, Pashin, 2007, Pashin, 2008, Pashin et al., 1991 and Scott, 2002).

CBM has been produced commercially in the Black Warrior Basin of Alabama since 1980, with cumulative production exceeding 69 $\times$ 10^9 sm^3 of gas and 1.6 $\times$ 10^9 bbl of water. In 2011, more than 5500 wells were active, producing more than 2.7 $\times$ 10^9 sm^3 of gas and 69 $\times$ 10^6 bbl of water. Analysis of production data indicates that a typical CBM well in the Black Warrior Basin produces cumulatively between 3 $\times$ 10^6 and 14 $\times$ 10^6 sm^3 of gas and 26,000 and 300,000 bbl of water (Pashin, 2010a). Initial research efforts in the 1970s focused on mitigating methane hazards in underground coal mines (Elder and Deul, 1974). The potential for production from multiple seams outside of mining areas was proven by the mid 1980s (Graves et al., 1983). With a rich history of production and reservoir data, the Black Warrior is an exceptional basin to explore the interrelationships among reservoir geology, water chemistry, and gas composition.

This study employs a variety of geological and geochemical techniques to characterize these interrelationships along a corridor from Brookwood Field, which is dominated by fresh formation water, to Blue Creek Field, where saline formation water is preserved in the interior of the basin (Fig. 1). This research is part of a multi-year study that is sponsored by the U.S. Department of Energy and is focused on the sampling and characterization of reservoir fluids in CBM reservoirs of the Black Warrior Basin. The primary goals of this research are to increase understanding of water and gas chemistry in CBM reservoirs and to use this understanding to develop improved produced water and reservoir management strategies.

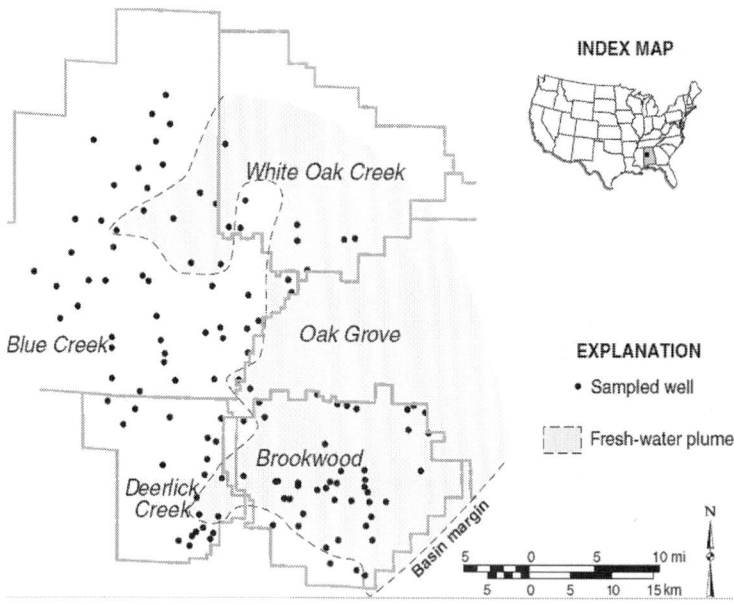

Figure 1: Map of study area in the Black Warrior basin of Alabama showing locations of wellhead samples in Brookwood, Oak Grove, White Oak Creek, Blue Creek, and Deerlick Creek fields.

ANALYTICAL METHODS

This research draws on a database that includes 91 analyses of the geochemistry of produced water. These data include physical and aggregate water properties, as well as the concentrations of major ionic compounds, metallic and nonmetallic constituents, and organic constituents. Produced water samples were collected between 2010 and 2012 at wellheads. Field parameters (conductivity, pH., turbidity) were recorded using a Horiba U50 Multi Water Quality Checker. Raw water was collected in the instrument's sample vessel, and the instrument was inserted into the vessel for analysis. The tool was calibrated using Horiba calibration fluid, distilled water, and pH 4 and 7 standard solutions before each measurement. Samples were prepared for laboratory analysis at the Geological Survey of Alabama, the University of Alabama, and the U.S. Geological Survey (USGS). Raw water was collected in two 250 ml Whirl-Pak bags as backup and to determine TSS. For filtered samples, water was passed through glass fiber prefilters and 0.45-μm filter membranes using a plastic-vacuum-hand pump. The filtered water was decanted into a series of Whirl-Pak bags, polyethylene bottles, and glass bottles. A series of unfiltered samples also was collected in a similar set of receptacles. Some receptacles and samples were treated with H_2SO_4, NaOH, and HNO_3 as prescribed by standard sampling procedures for the various analytes. All samples were chilled to ~ 4°C in coolers for transport to the laboratory.

In the laboratory, a Dionex 4000i ion chromatograph was used to determine the concentrations of Br, F, NO_3, PO_4, and SO_4 with precision exceeding 0.06 mg/L. A Perkin-Elmer 3000 DV ICP-MS was used to determine many metallic and nonmetallic constituents, including Ag, Al, B., Ba, Be, Ca, Cd, Co, Cr, Cu, Fe, K, Li, Mg, Mn, Mo, Na, Ni, SiO_2, Sn, Sr, Ti, V, and Zn, with precision varying between 1 μg/L and 1 mg/L depending on the analyte. A Perkin-Elmer 5100PC GFAAS was used to quantify key trace elements (As, Cs., Pb, Rb, Sb, Se, and Ti) with precision exceeding 2.0 μg/L, and a Perkin-Elmer 2380 CVAAS was used to determine Hg with a sensitivity of detection of 0.010 μg/L. TOC was determined using a Shimadzu TOC-5000A analyzer. Alkalinity, Cl, CN, F, NH_3, NO_2, P, phenolic compounds, and TKN were determined using a Technicon Autoanalyzer. All determinations were made using third-party quality control samples and using standard procedures specified by the U.S. Environmental Protection Agency and the U.S. Geological Survey. For quality control, replicate analyses were run on every 10th sample, and ionic concentrations were checked for charge balance. Reproducibility of measurements was high, with an error of < 2% for most analytes. Additional samples were collected and processed by the U.S. Geological Survey for GC-MS analysis of the organic compounds and selected inorganic compounds in the produced water, including DOC, extractable hydrocarbons, volatile fatty acids, PO_4, and NH_4 using the analytical methods described by Orem et al. (2007) and in a supplemental file that is included with this publication.

Wellhead gas samples were collected from 56 well sites using an Isotube sampling system. The samples were sent to Weatherford Laboratories for geochemical analysis. Analysis of gas composition included the determination of the quantities of hydrocarbon (C_1–C_6) and nonhydrocarbon gases (He, Ar, O_2, N_2, CO_2). Composition was determined by gas chromatograph with analytical precision of +/– 5% for most gases and of +/– 10% for C_{4-6} hydrocarbons. Oxygen concentrations were used to correct the analyses for atmospheric contamination. Stable isotope analysis of methane was performed on the gas samples and included $^{13}C_1$ and D determinations. Isotopic determinations were made on a GC-C-IRMS unit with analytical precision of 0.3‰ for $^{13}C_1$ and 3.0‰ for D.

Geochemical data from the water and gas analyses were compiled in a spreadsheet and incorporated into a Petra database that contains a broad range of geological information, including well locations, stratigraphic data, and structural data. Bubble maps were made in Petra that show variation of water chemistry and gas composition in the study area. Major ionic concentrations were plotted on Piper diagrams to classify the formation water, and stable isotopic data were cross-plotted to help interpret the origin of the coalbed gas. Compositional data also were plotted against other geological and geochemical variables, such as coal rank parameters and TDS, to determine basic geologic controls on water and gas composition. Summary tables are included in this paper, and supplemental files containing the basic data are available online alongside the PDF of this paper through http://www.sciencedirect.com.

Mineral cements provide valuable information on the geochemical evolution of sedimentary basins, including hydrodynamic and methanogenetic processes in organic-rich strata (e.g., Budai et al., 2002 and Pitman et al., 2003). Fracture-filling calcite was extracted from 22 long cores distributed among the Black Warrior CBM fields. Samples were extracted by hand and by drill from veins (i.e., joint and fault-related fracture fills) in sandstone and shale and from cleats in coal. Stable isotopic analysis was performed to determine ^{13}C and ^{18}O values for the calcite cement. Isotopic composition was measured with a GasBench-IRMS system using a method similar to that described by Debajyoti and Skrzypek (2007). Isotope values are expressed relative to the VPDB scale by use of the NBS-19 standard.

GEOLOGICAL FRAMEWORK

The Black Warrior Basin is a late Paleozoic foreland basin in Alabama and Mississippi (Thomas, 1985 and Thomas, 1995). The CBM fields are situated near the southeastern margin of the basin along the frontal structures of the Appalachian orogenic belt (e.g., Pashin, 2007, Pashin, 2010b and Pashin and Groshong, 1998) (Fig. 2). Economic coal and CBM resources are in Upper Carboniferous (Lower Pennsylvanian) strata of the Pottsville Formation. Coal seams are distributed through a stratigraphic section that is generally between 0.5 and 1.3 km thick and is composed principally of shale and sandstone. Seam thickness ranges from less than 0.1 m to more than 3.0 m, and beds as thin as 0.3 m are routinely completed for production. Five to eight seams have been completed in most wells within the study area, and virtually all wells have been hydrofractured in multiple stages. Typical completion depths within this area are between 200 and 750 m, and vertical wells are typically developed with a 16–32 ha (40–80 acres) spacing. Horizontal and gob wells are drilled in conjunction with deep longwall coal mining operations in Brookwood and Oak Grove fields, and these mine-related wells are not part of this study.

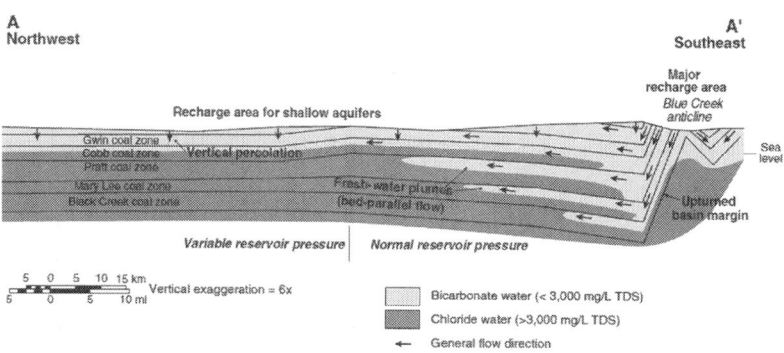

Figure 2: Structural cross-section showing fresh-water recharge area along the structurally upturned, southeastern margin of the Black Warrior basin (after Pashin et al., 2004).

Strata in the eastern Black Warrior Basin have regional dip of less than 1° toward the southwest. Folds of the Appalachian thrust belt have deformed the southeastern part of the regional dipping panel. An Appalachian fold limb along the southeastern edge of the basin forms an upturned basin margin that strikes about N. 50° E. The major reservoir coal beds that are productive in the interior of the basin are exposed at the surface in the fold limb (Fig. 2). This upturned basin margin is a key geologic feature that has a strong influence on basin hydrology (Pashin and McIntyre, 2003 and Pashin et al., 1991). Pottsville strata in the CBM fields are offset by a multitude of high-angle normal faults that strike northwest and define a horst-and-graben system with net displacement increasing toward the southwest (Pashin and Groshong, 1998, Pashin et al., 2004 and Thomas, 1988).

Coal in the Pottsville Formation is bright-banded and ranges in rank from high volatile B bituminous to low volatile bituminous in the CBM fields (Levine and Telle, 1989, Pashin et al., 1999, Telle et al., 1987, Winston, 1990a and Winston, 1990b). The Mary Lee coal zone is of high volatile B bituminous rank in the western part of the study area, and rank increases toward the east and southeast (Fig. 3). McFall et al. (1986) were first to recognize that cleat spacing tends to decrease as coal rank increases in the Black Warrior Basin. High volatile A bituminous coal forms a broad belt extending from

Deerlick Creek Field northward through Blue Creek and White Oak Creek fields. A large area containing medium and low volatile bituminous coal is developed in Oak Grove and Brookwood fields and forms the heart of Alabama›s metallurgical coal industry. Isovols cross-cut structure, indicating that the rank pattern is post-kinematic and reflects regional variation of geothermal gradient rather than simple burial depth (Pashin et al., 1999 and Telle et al., 1987). Explanations of the rank pattern include thermal upgrading from orogenic expulsion of warm fluid (Winston, 1990a and Winston, 1990b) and increased overburden related to an eroded thrust sheet (Thomas et al., 2008). Analysis of burial history indicates that major thermal maturation occurred near maximum burial during late Paleozoic orogenesis (Carroll et al., 1995 and Telle et al., 1987) (Fig. 4).

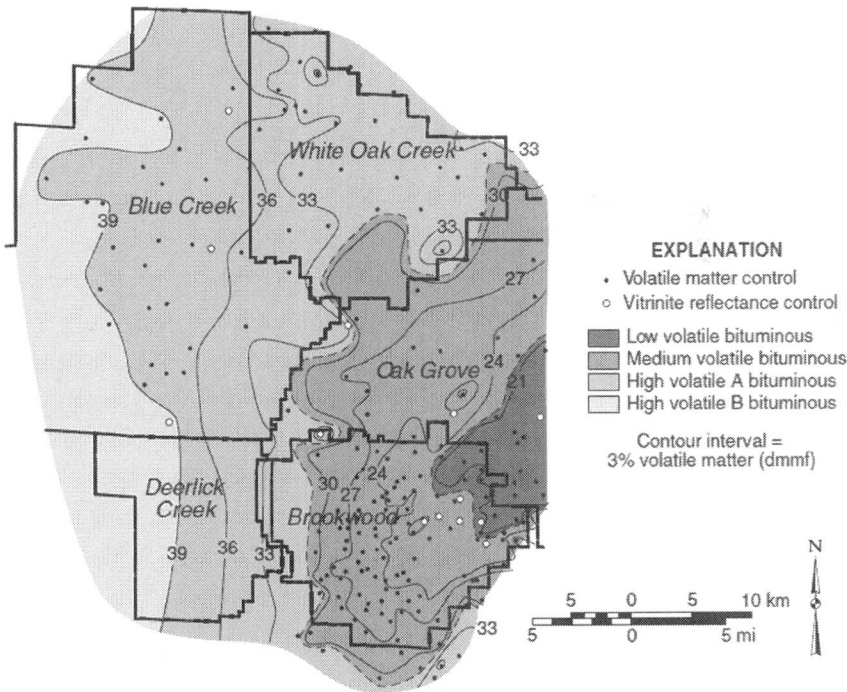

Figure 3: Coal rank in the study area based on volatile matter (dmmf) in the Mary Lee coal zone.

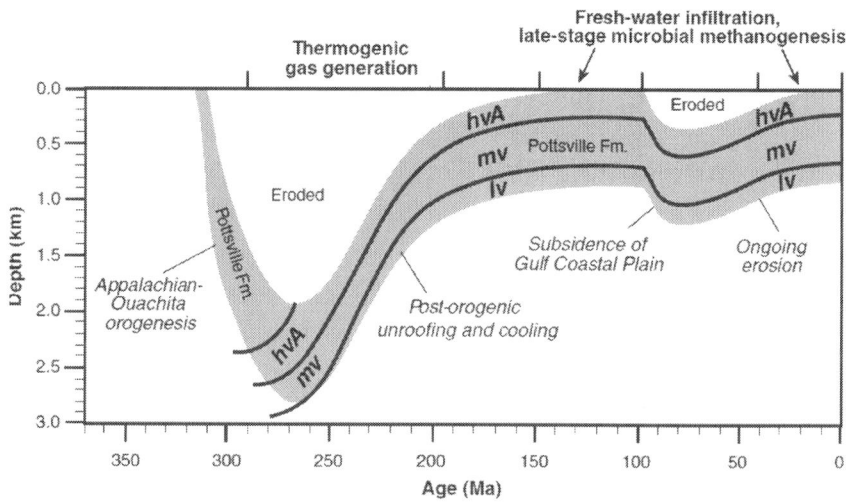

Figure 4: Burial and thermal maturation history of the eastern Black Warrior basin (after Pitman et al., 2003).

Shale, sandstone, and coal within the CBM target interval have minimal matrix permeability, and so subsurface flow is concentrated in natural fractures. Cleats in Pottsville coal are typically spaced between 0.1 and 2.5 cm. These closely spaced fractures enable commercial flow rates in coal. Bounding shale and sandstone units function as confining beds (Pashin et al., 2004). Coal is a stress-sensitive rock type, and so as overburden stress increases, cleat aperture decreases, thereby decreasing permeability exponentially from about 1 D near the surface to less than 1 mD at depths beyond 700 m (McKee et al., 1988). Consequently, most water sampled at wellheads is recovered primarily from shallow, permeable coal seams. Water in the Pottsville Formation varies from fresh in the eastern part of the study area to saline brine in the western part (Ellard et al., 1992 and Pashin et al., 1991). Water with less than 3000 mg/L total dissolved solids (TDS) is present at reservoir depth adjacent to the upturned basin margin (Fig. 2). By contrast, TDS content locally exceeds 60,000 mg/L to the west. The fresh water is thought to be the product of meteoric recharge of coal along the basin margin (Pashin, 2007 and Pashin et al., 1991). Extension of fresh-water plumes into the interior of the basin is interpreted to support reservoir pressure and has a significant impact on the saturation and mobility of coalbed gas (Pashin, 2010b).

WATER CHEMISTRY

Total Dissolved Solids and Major Ionic Compounds

The composition of water produced from Black Warrior CBM reservoirs varies from fresh and brackish water that is protected by the Safe Drinking Water Act (< 10,000 mg/L TDS) to basinal brine (TDS > 30,000 mg/L) (Fig. 5). TDS content within the study area ranges from 606 mg/L to more than 61,000 mg/L, and major ionic compounds are dominated by Cl^-, Na^+, and HCO_3^- (Table 1). The produced water represents a compositional continuum ranging from $NaHCO_3^-$ to NaCl-type, which is common in many CBM plays (Ayers and Kaiser, 1994; C.A. Rice et al., 2000; C.A. Rice, 2003 and Van Voast, 2003). A few samples contain anomalous concentrations of Ca^{2+}, Mg^{2+}, and SO_4^{2-}, and all of these samples have TDS < 10,000 mg/L. Regression analysis indicates that Cl^- and Na^+ are the principal determinants of TDS content (Fig. 5). Bicarbonate (HCO_3^-) content correlates negatively with TDS. To date, the maximum recorded HCO_3^- content in water with TDS > 10,000 mg/L is 429 mg/L, and NaCl-type water predominates where TDS > 3000 mg/L.

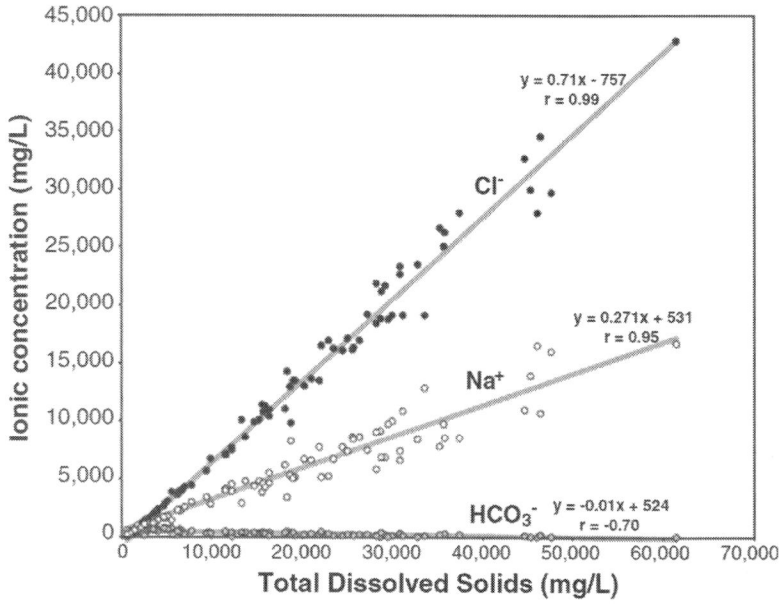

Figure 5: Plot of major ionic compounds versus TDS content in Black Warrior CBM reservoirs.

Table 1: Summary of major ionic compounds and selected physical and aggregate properties of wellhead water samples from coalbed methane wells in the Black Warrior Basin. Analyses performed in the geochemical laboratories of the Geological Survey of Alabama and the University of Alabama

Variable	N	Minimum	Maximum	Mean	Standard deviation
HCO_3^- (mg/L)	91	2	978	347	205
Ca^{2+} (mg/L)	91	2	1350	231	270
Cl^- (mg/L)	91	0	42,800	11,198	9865
K^+ (mg/L)	91	1	74	15	14
Mg^{2+} (mg/L)	91	0	392	75	83
Na^+ (mg/L)	91	236	16,700	5103	3941
SO_4^{2-} (mg/L)	91	0	302	6	35

Total dissolved solids (mg/L)	91	606	61,374	16,846	13,766
Total suspended solids (mg/L)	91	0	1020	77	118
pH	90	5.3	9.0	7.5	0.6
Turbidity (NTU)	90	2	412	94	88

Mapping TDS defines the fresh-water plumes (TDS < 10,000 mg/L) and saline basinal fluid (TDS > 30,000 mg/L) (Fig. 6). Mapping HCO_3^- levels establish an inverse pattern (Fig. 7), reflecting the meteoric origin of the fresh-water plumes, in which HCO_3^- enrichment is a by-product of biological activity (e.g., Van Voast, 2003). In Brookwood Field, 6 samples within the plumes have TDS > 10,000 mg/L (Fig. 6). These pockets of saline water may reflect local sheltering from the regional recharge system by small-displacement normal faults (Pashin et al., 1991). Conversely, isolated samples basinward of the plumes with < 10,000 mg/L TDS may reflect local extensions of the plumes or, alternatively, percolation along faults and fracture zones.

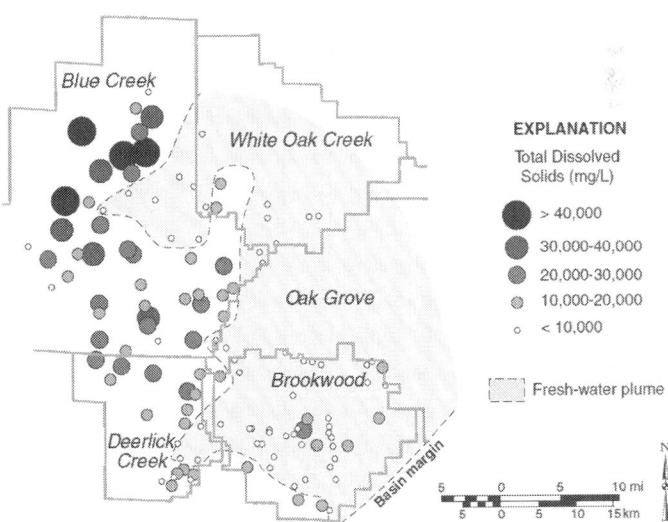

Figure 6: Map of TDS content showing fresh water in the eastern part of the study area and saline to hypersaline water in the northwestern part.

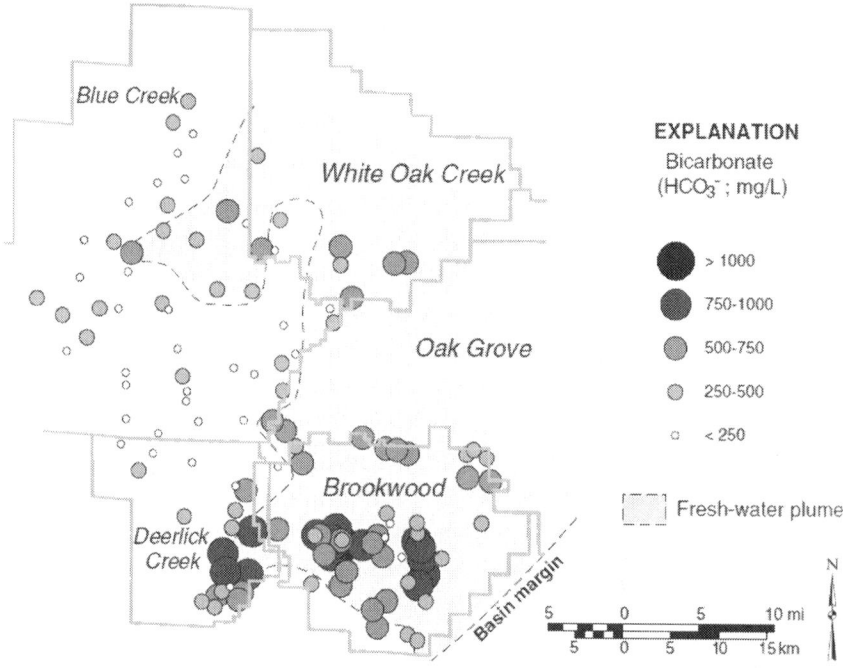

Figure 7: Map of bicarbonate content showing highest concentrations within the fresh-water plumes in the eastern part of the study area.

Metallic and Nonmetallic Constituents

Metallic and nonmetallic substances tend to be minor components of the produced water (Table 2 and Table 3), and many correlate positively with chloride. Phosphorus and phosphate concentrations are typically lower than 500 µg/L but are locally higher than 3000 µg/L and do not correlate with each other or other geochemical variables. Dissolved SiO_2 concentrations range from 1 to 18 mg/L. Dissolved CO_2 concentrations average at 37 mg/L and locally exceed 600 mg/L. Values higher than 100 mg/L are restricted to water with TDS < 10,000 mg/L.

Table 2: Summary of metallic and nonmetallic elements in wellhead water samples from coalbed methane wells in the Black Warrior Basin. Analyses performed in the geochemical laboratories of the Geological Survey of Alabama and the University of Alabama

Substance	N	Minimum(μg/L)	Maximum(μg/L)	Mean(μg/L)	Standard deviation(μg/L)
Al	91	11	99	42	19
As	91	0	12	1	2
B	91	0	493	177	123
Ba	91	136	281,000	52,230	54,360
Be	91	0	1	0	0
Br	91	0	150	46	36
Cd	91	0	12	1	2
Co	91	0	115	24	24
F	91	582	22,600	11,345	6123
Fe	91	70	93,100	12,833	18,259
Hg	91	0	0	0	0
Li	91	0	8940	1356	1510
Mn	91	6	4840	351	566
Mo	91	0	67	2	8
Ni	91	0	358	22	46
P	91	0	3250	325	468
Pb	91	0	155	6	18
Sb	91	0	22	0	2
Se	91	0	41	2	7
Sn	91	0	9	0	1
Sr	91	45	142,000	13,008	19,645
Ti	91	0	39	3	7
V	91	0	39	1	6
Zn	91	0	278	26	40

Table 3: Summary of nutrients, organic compounds, and other constituents in wellhead water samples from coalbed methane wells in the Black Warrior Basin. Analyses performed in the geochemical laboratories of the Geological Survey of Alabama, the University of Alabama, and the USGS

	N	Minimum	Maximum	Mean	Standard deviation
CO_3 (mg/L)	91	0	64	2	7
CO_2 (mg/L)	90	0	634	37	93
NH_3 and NH_4^+ as N (mg/L)	91	0	24	5	4
NH^{4+} (mg/L)	58	0	9	4	3
NO_2 as N (mg/L)	91	0	0.63	0.02	0.09
NO_3 as N (mg/L)	91	0	28	2	5
PO_4 (µg/L)	58	26	3570	435	554
SiO_2 (mg/L)	91	1	18	8	3
DOC (mg/L)	55	1	61	3	9
TOC (mg/L)	91	0	85	7	15
Aromatic hydrocarbons[a] (µg/L)	91	0	103	17	20

[a]Reported as phenolic compounds.

Trace elements of note are Br, Co, Ni, Pb, Se, V, and Zn. Of these, Br, Co, Ni, and Zn correlate positively with chloride and are thus associated with basinal brine. Mercury (Hg) was consistently below detection levels (0.001 µg/L). The amount of Pb is lower than 20 µg/L in all but 5 samples, which have anomalous Pb concentrations at TDS > 18,000 mg/L. Selenium is typically below detection levels (1 µg/L), but 5 samples have concentrations above 20 µg/L where TDS > 30,000 mg/L. Similarly, V is below detection levels (4 µg/L) in all 4 samples. In those 4 samples, V is above 14 µg/L, and TDS > 30,000 mg/L.

Nitrogen Compounds

Nitrate (NO_3^-), ammonia, and ammonium (NH_3 and NH_4^+ determined as N) are the principal nitrogen-bearing compounds in the produced water (Table 3). Nitrate levels are characteristically lower than 19.3 mg/L and do not correlate with other geochemical parameters. However anomalously high NO_3^- values (36–123 mg/L) were measured in 9 samples of saline formation water (Fig. 8). Nitrite (NO_2^-) concentrations in formation water are typically below detection limits. Where detected, levels are lower than 0.7 mg/L. Ammonia and ammonium, by comparison, were detected in all samples, and values range from less than 1 to 24 mg/L. Ammonium constitutes about 60% of the nitrogen compounds in the water. Ammonia + ammonium content correlates strongly and positively with TDS content (Fig. 9), and concentrations higher than 5 mg/L are typical of water with TDS > 10,000 mg/L (Fig. 10).

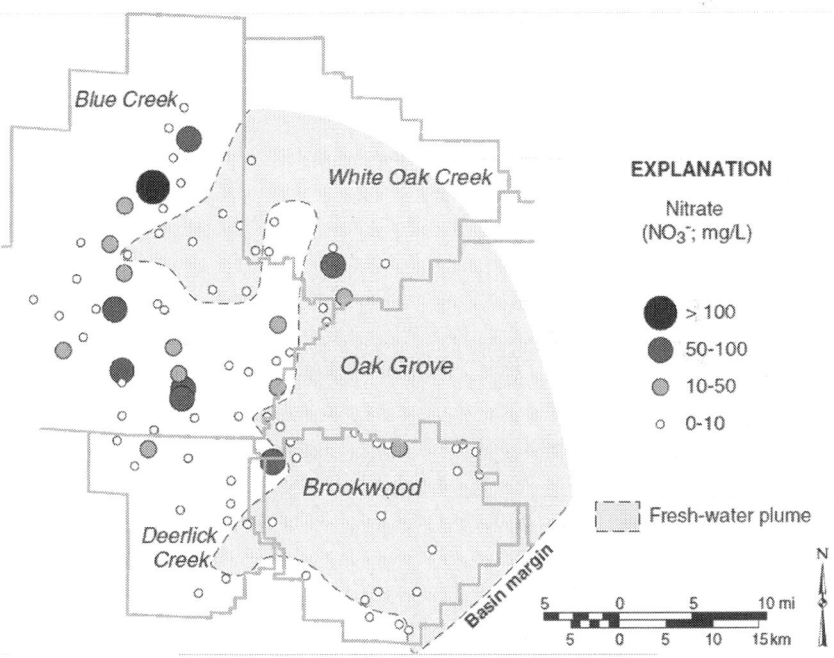

Figure 8: Map of nitrate content showing greatest concentrations in the northwestern part of the study area.

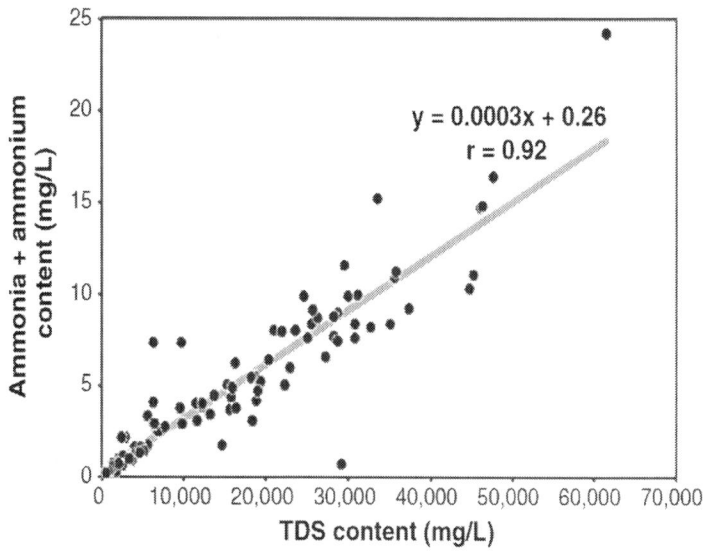

Figure 9: Plot of ammonia + ammonium content (as N) versus TDS content showing strong positive correlation.

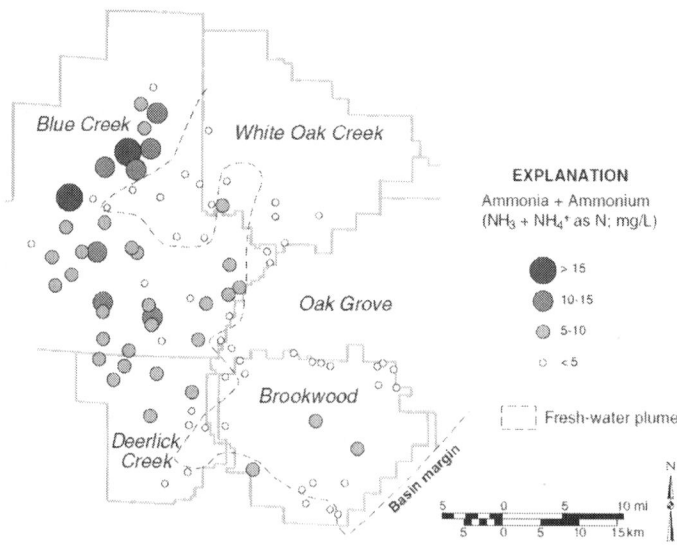

Figure 10: Map of ammonium + ammonia content (as N) showing greatest concentrations in the northwestern part of the study area.

The clear correlation between $NH_3 + NH_4^+$ and TDS (Fig. 9) indicates a strong association with basinal brine, which is surprising for these compounds. Ammonium is known to substitute for K and Na in feldspar and muscovite in metamorphic terranes (Barker, 1964, Duit et al., 1986 and Eugster and Munoz, 1966). Ammonia, moreover, has been identified in illite from Carboniferous coal-bearing strata (Juster et al., 1987) and in oil shale (Oh et al., 1988). In the North German Basin, NH_4^+-bearing illite is thought to be a source of N_2 in natural gas (Mingram et al., 2005). Therefore, one explanation for the correlation between $NH_3 + NH_4^+$ and TDS is ion exchange between minerals and brine. However, Bates et al. (2011) attributed NH_4^+ in water produced from subbituminous coal of the Powder River Basin to reaction of formation water with coal. One possibility is that denitrification of organic matter early in the regional coalification history may be the ultimate source of NH_4^+ and NH_3 in the Black Warrior Basin, and the distinct association with basinal brine is interpreted to reflect a complex history of nitrogen exchange among organic matter, minerals, and formation water.

Organic Constituents

DOC and TOC values in the produced water are typically < 12 mg/L but locally exceed 60 mg/L (Table 3). A variety of organic compounds were identified, and the produced water has a distinct petroliferous odor. Organic compounds include a variety of aromatic, heterocyclic, and aliphatic hydrocarbons. Indeed, several of the wells from the northwestern part of the study area produce incidental amounts of oil (< 1 bbl/day). Concentrations of aromatic hydrocarbons in the produced water are locally higher than 100 µg/L (Fig. 11; Table 3). Phenols and polycyclic aromatic hydrocarbons (PAH) form the bulk of the extractable hydrocarbons, which represent a fraction of the total hydrocarbons in the formation water (Table 4). Additional extractable hydrocarbons include minor amounts of other heterocyclic, aromatic, and non-aromatic compounds. Overall, the assemblage and concentrations of organic compounds in Black Warrior coal resemble those identified previously in coal-borne water of the Powder River Basin (Orem et al., 2007). The dominant phenols that were identified are 4-(1,1,3,3-tetramethylbutyl)-phenol and 2,4-bis(1,1-dimethylethyl)-phenol. The principal PAH compounds are naphthalic substances. Acetate was the only volatile fatty acid identified, and concentrations are at < 1 mg/L. Other fatty acids, including a variety of decanoic and decenoic acids, were detected in the produced water.

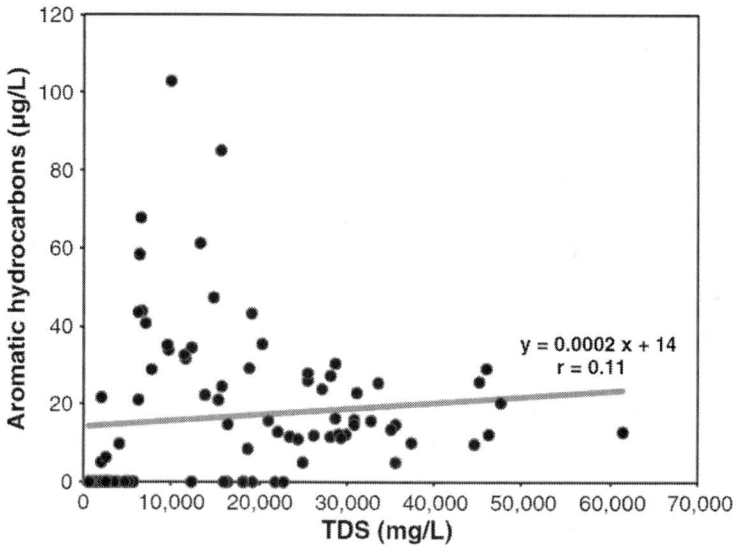

Figure 11: Plot of aromatic hydrocarbon content versus TDS in produced water.

Table 4: Summary of selected organic compounds with maximum concentrations greater than 2.00 μg/L from extracts of produced water samples from the Black Warrior basin. Samples analyzed in the geochemical laboratories of the U.S. Geological Survey

Compound	Wells with detections (N/65)	Estimated concentration range(μg/L)
Polycyclic aromatic hydrocarbons		

Naphthalene	49	0.01–6.57
Methyl-naphthalene	52	0.01–15.55
Dimethyl-naphthalene	39	0.01–9.51
Trimethyl-naphthalene	23	0.01–4.49
Methyl-biphenyl	18	0.01–2.13
Heterocyclic compounds		
Methyl-quinoline	31	0.03–3.75
Benzothiazole	45	0.01–3.04
Caprolactam	10	0.02–2.39
Phenols		
4-(1,1,3,3-tetramethylbutyl)-phenol	17	0.01–18.34
2,4-bis(1,1-dimethylethyl)-phenol	21	0.01–4.94
Other aromatics		
Dioctyl phthalate	57	0.01–2.30
Non-aromatic compounds		
Triphenyl phosphate	6	0.01–6.77
Tributyl phosphate	23	0.01–2.66
Cyclic octaatomic sulfur	29	0.10–9.63
Dodecanoic acid	30	0.67–2.52
Tetradecanoic acid	53	0.94–5.32
Hexadecanoic acid	50	1.17–3.02
Octadecanoic acid	32	1.62–3.73
Hexadecenoic acid	25	1.13–8.37
Octadecenoic acid	29	1.60–3.40

Total aromatic hydrocarbon concentration, reported as phenolic compounds in the Autoanalyzer results, has a complex relationship to TDS (Fig. 11). The lower limit of detection is 3 µg/L, and samples below the detection limit are from areas where TDS < 22,000 mg/L. Concentrations < 30 µg/L are common across all recorded TDS values, and elevated concentrations of 30 to 103 µg/L form a peak between TDS levels of 6000 and 20,000 mg/L. Samples with aromatic hydrocarbon constituents exceeding 25 µg/L come primarily from areas containing high volatile A and high volatile B bituminous coal (Fig. 12), which lie in the main oil generation window. Hence, a logical conclusion is that the hydrocarbons are associated with the catagenesis of coal. Following this line of reasoning, low concentrations in medium volatile bituminous and low volatile bituminous coal, which are in the heart of the thermogenic gas window, could be interpreted as a product of thermal alteration.

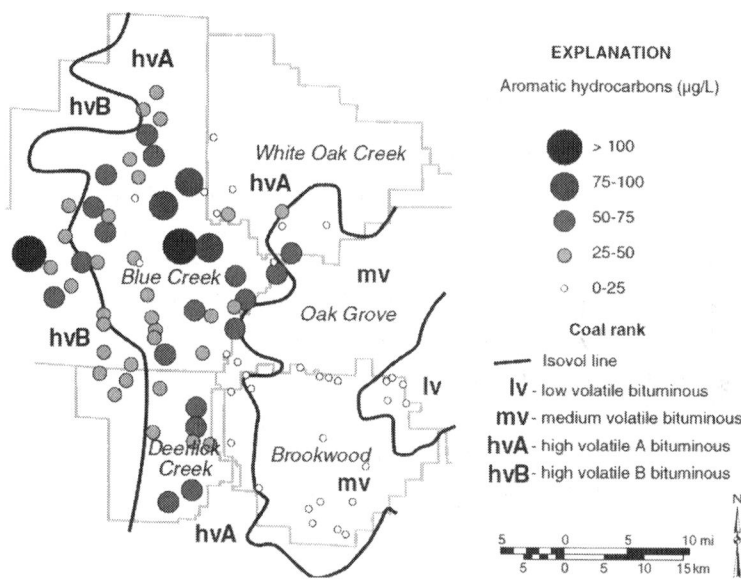

Figure 12: Map of aromatic hydrocarbon content showing elevated concentrations in high volatile A and high volatile B bituminous coal.

Alternatively, the similarity of the extractable hydrocarbons to those identified in thermally submature coal of the Powder River Basin (Orem et al., 2007) suggests that many of these compounds formed either prior to catagenesis or, perhaps preferably, after catagenesis was complete. Fatty acids, for example, may be by-products of late-stage microbial activity in the coal seams. Samples with concentrations below detection limits come primarily from Brookwood and White Oak Creek fields, suggesting that the distribution of hydrocarbon compounds reflects not only the regional rank pattern, but also may reflect basinward transport of hydrocarbons during fresh-water intrusion. Indeed, the peak of aromatic hydrocarbon concentrations between TDS levels of 6000 and 20,000 mg/L (Fig. 11) may indicate accumulation of flushed hydrocarbons in the distal reaches of the plumes.

GAS COMPOSITION

Bulk Composition

Wellhead gases produced from the Black Warrior CBM fields are dominated by CH_4 (Table 5). Bulk composition is remarkably uniform, and nonhydrocarbon gas content is typically less than 1%. Carbon dioxide concentrations are lower than 0.58% and average only at 0.16%, which is unusually low for CBM reservoirs (Scott, 1993). Indeed, significant volumes of CO_2 are typically generated thermogenically and biogenically from humic source rocks like coal (Bates et al., 2011, Hunt, 1979 and Scott et al., 1994). Note, however, that although little CO_2 is in the produced gas stream, a significant volume of CO_2 is dissolved in the formation water (Table 3).

Table 5: Summary of geochemical analyses of wellhead gas analyses from coalbed methane wells in the Black Warrior Basin. Analyses performed by Isotech and Weatherford Laboratories

Variable	N	Minimum	Maximum	Mean	Standard deviation
CO_2 (%)	56	0.023	0.580	0.159	0.092
N_2 (%)	51	0.008	5.801	0.727	0.931
CH_4 (%)	56	93.400	99.909	99.090	1.010
C_2H_6 (%)	56	0.005	0.848	0.076	0.124
C_3H_8 (%)	46	0.001	0.315	0.013	0.046
Iso-butane (%)	14	0.000	0.035	0.004	0.009
Butane (%)	15	0.000	0.049	0.004	0.012
Dryness index $100*(C_1/C_{1-5})$	56	98.7	100.0	99.9	0.2

$\delta^{13}C_1$ (‰)	56	– 60.37	– 42.28	– 51.16	4.04
δD_{C1} (‰)	56	– 206.0	– 185.7	– 197.8	3.5

Nitrogen is the most abundant nonhydrocarbon gas, with concentrations locally exceeding 5% (Table 5). The mean N_2 concentration is only 0.73%. Reported N_2 concentrations from the 1980s and 1990s are substantially higher (> 5%) (D.D. Rice, 1993 and Scott, 1993), and most of these samples came from the fresh-water plumes. Nitrogen tends to be produced early in the life of CBM wells because it is a much weaker adsorbate than CO_2 and CH_4 (e.g., Hall et al., 1994). Some of the N_2 may be derived from fixation of atmospheric gas in rain and stream water, as well as from soil gas in the recharge area. Other potential sources include thermal and microbial degradation of organic matter, mineral matter, and formation water, and this possibility is discussed in Section 5.2.

Mean CH_4 content averages more than 99%, and the standard deviation is only 1%. The total volume of wet hydrocarbon gases (C_2–C_5) is less than 0.1% of total gas volume. Indeed, C_4 hydrocarbons (butane and iso-butane) were detected in only 15 samples, and C_5 hydrocarbons were not detected. The produced gas is exceptionally dry, with a mean dryness index of 99.9 and a standard deviation of only 0.2. The consistent gas composition, low concentrations of nonhydrocarbon gases, and absence of H_2S facilitate gas production at pipeline quality. Thus, dehydration and compression constitute all the processing that is required to deliver Black Warrior CBM to market.

Stable Isotopes

Although bulk gas composition varies little across the study area, stable isotopic composition varies considerably and is of value for characterizing the origin of the gas. Values of $\delta^{13}C_1$ range from – 60.37 to – 42.28‰ and have a weak central tendency (Table 5). By contrast, δD_{C1} values range narrowly from – 185.7 to – 206.0‰ and have a strong central tendency around a mean value of – 197.8‰.

Cross-plotting $\delta^{13}C_1$ and δD_{C1} values on the diagram of Whiticar (1996) indicate the data cluster in the thermogenic and mixed thermogenic–biogenic generation fields (Fig. 13). The samples that are most depleted in ^{13}C plot at the edge of the biogenic gas field. The D_{C1} values suggest that microbial CO_2 reduction was the principal metabolic pathway for biogenesis, although this interpretation is made with caution because exchange of H between CH_4 and formation water can result in misleading δD values (Vinson et al., 2012).

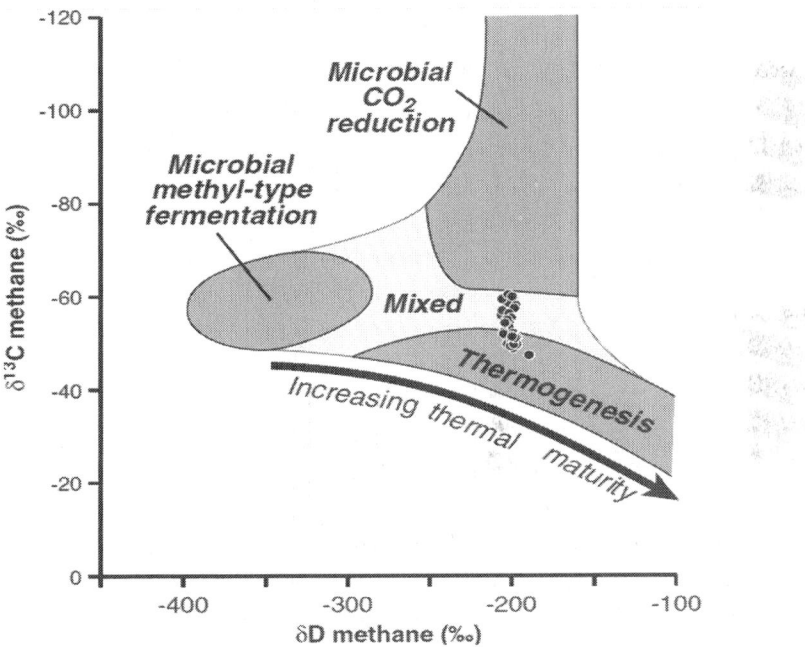

Figure 13: Cross-plot of $\delta^{13}C$ versus δD for methane from wellhead gas samples. Results suggest mixing of thermogenic gas with late-stage bio-genic gas that formed along a CO_2 reduction metabolic pathway.

Importantly, microbial gas generated by CO_2 reduction can have $\delta^{13}C_1$ values substantially greater than − 55.0‰ depending on the source CO_2 (Flores et al., 2008, Martini et al., 2003, Martini et al., 2008 and Whiticar, 1999), and so there is no precise cutoff value that distinguishes mixed gas from pure biogenic gas. Basins like the

Black Warrior, which have been uplifted and cooled substantially, are predicted to be significantly undersaturated with thermogenic gas (Bustin and Bustin, 2008 and Scott, 2002), and so a significant component of microbial gas helps explain the high gas saturation that has been observed in many Pottsville coal seams (Pashin, 2007 and Pashin, 2010b). Microbial CO_2 reduction involves the conversion of CO_2 to CH_4 in the presence of H_2 (e.g., Whiticar et al., 1986). The basic reaction can be expressed as

$$CO_2 + 4H_2 \rightarrow CH_4 + 2H_2O.$$

(1)

Considering the exceptionally low percentage of CO_2 in the produced gas (Table 5), CO_2 dissolved in formation water constitutes the main resource that is available to drive CO_2 reduction (Table 3). Obvious sources of H_2 exist in the organic framework of the coal and the organic compounds in the formation fluids, including long-chain fatty acids and phenolic compounds, which have been degraded by microbes in biogenesis experiments (Jones et al., 2010). Plotting $\delta^{13}C_1$ against vitrinite reflectance of the Mary Lee coal zone demonstrates a significant correlation between thermal maturity and isotopic composition (Fig. 14). Comparison of isotopic data with mapped rank patterns further supports covariance of $\delta^{13}C_1$ values and thermal maturity, although there is considerable local variability in ^{13}C enrichment levels (Fig. 15). Commingling of gas from multiple coal seams, differences in the intensity of microbially driven isotopic fractionation, and local gas migration help explain these variations.

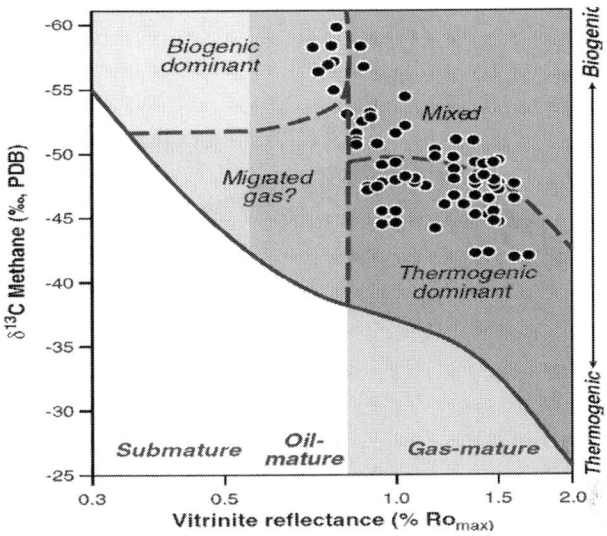

Figure 14: Plot of $\delta^{13}C$ ratios in methane versus the vitrinite reflectance of coal demonstrating correlation between stable isotopic composition and thermal maturity.

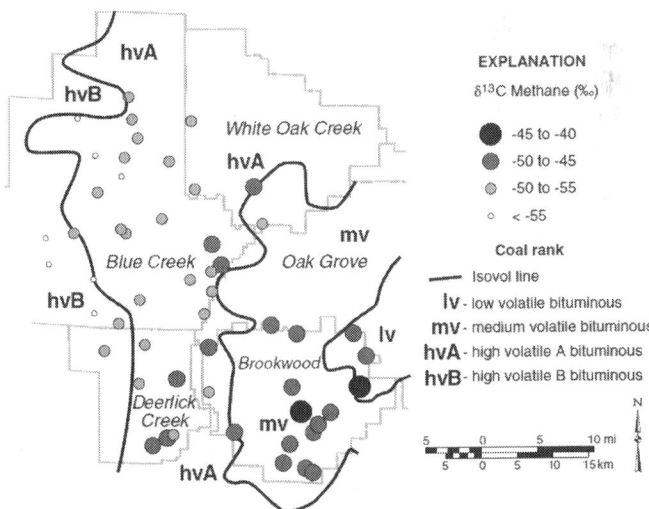

Figure 15: Map of $\delta^{13}C$ ratios in methane showing progressive enrichment of ^{13}C with increasing coal rank.

The lowest $\delta^{13}C_1$ values plot near the high volatile A–B bituminous rank transition, which corresponds to the edge of the thermogenic gas window for coal (Jüntgen and Karweil, 1966 and Jüntgen and Klein, 1975). Pashin (2007) noted that CH_4 produced from coal of medium and low volatile bituminous rank (R_o = 1.1–1.9; Taylor et al., 1998) in the Black Warrior Basin is more depleted in ^{13}C than that produced from coal of similar thermal maturity in other basins. The highest rank coal is in the fresh-water plumes close to the recharge area, where intense microbial activity would be predicted. In this type of setting, however, intense reduction of CO_2 may actually enrich CH_4 in ^{13}C, as has been observed in the San Juan and Powder River basins (Bates et al., 2011 and Scott et al., 1994). Because of this, distinguishing the relative impact of thermogenesis and biogenesis in coal that has entered the gas window and contains fresh water is difficult without additional information on inorganic carbon in CO_2, formation water, or mineral cement, the latter of which is discussed below in Section 6.

Nitrogen compounds, including NH_3, NH_4^+, NO_2^-, and NO_3^-, are important nutrients that are integral parts of the nitrogen cycle (e.g., Galloway et al., 2004 and Karl and Michaels, 2001), but relatively little is known about the role of these compounds in unconventional reservoirs. Indeed, the presence of these compounds in Pottsville formation water raises questions about how these substances may influence subsurface gas composition. For example, deionization of ammonium, expressed as

$$2NH_4^+ \rightarrow 2NH_3 + H_2.$$

$$(2)$$

could provide some of the H_2 for microbial CO_2 reduction while simultaneously explaining the NH_3 in the formation water. The low level of NO_2^-, which is a fouling agent that inhibits biological colonization of subaqueous and subsurface environments, is another factor that probably helped sustain microbial communities. Biologically mediated anaerobic oxidation of NH_4^+, a.k.a. the anammox process (Thamdrup and Dalsgaard, 2002 and Van de Graaf et al., 1995), may be a significant natural process in Black

Warrior CBM reservoirs. The relevant anammox reaction is:

$$NH_4^+ + NO_2^- \rightarrow N_2 + 2H_2O.$$ (3)

Although this specific process has yet to be substantiated in the deep subsurface, the implications of this type of reaction for late-stage biogenic gas generation in Pottsville CBM reservoirs are twofold. First, this reaction consumes NO_2^-, thereby inhibiting fouling of the reservoir. Second, this and a host of other reactions could account for a significant portion of the N_2 in the coalbed gas.

Assessing the sources of N_2 in the produced gas is difficult because an unknown amount of the parent material may have been consumed. Material balance provides a rudimentary level of constraint and a sense of scale based on the solubility of N_2 in water and the concentration of the remaining nitrogen compounds in the formation water. At 30 °C, the solubility of N_2 in water is ~ 50 mg/L, which equates to 6850 $sm^3/10^6$ bbl. The remaining nitrogen compounds, by comparison, have equivalent N_2 concentrations on the order of 7 mg/L, which equates to a generative potential of 1055 $sm^3/10^6$ bbl. A typical Black Warrior CBM well has a cumulative production of about 0.1×10^6 bbl of water and 8.5×10^6 sm^3 of gas (Pashin, 2010a). Of this, N_2 accounts for about 0.7%, or 59×10^3 sm^3 of the produced gas. The intrusion of fresh water at ~ 24 km into the basin suggests that > 30 pore volumes of water have swept through a given 16 ha (40-acre) drilling unit over geologic time. If 50% of a pore volume has been recovered, the produced water constitutes 1.67% of the water that has passed through the reservoir over geologic time. Thus, the N_2 derived from meteoric sources, such as the atmosphere and soil gas, would account for ~ 70% of the N_2 in the natural gas, and denitrification of dissolved compounds based on current concentrations would account for ~ 10%. By this logic, thermal and microbial degradation of other sources, such as coal and mineral matter, would account for the remaining 20%. Of course, geochemical and hydrodynamic variables change greatly across the basin. Accordingly, N_2 derived from meteoric sources is expected to be most abundant near the recharge zone, whereas the proportion of N_2 derived from organic and inorganic compounds is expected to increase toward the interior of the basin. In sum, nitrogen appears to be an important component of the CBM system in the Black Warrior Basin, and the role of nitrogen in unconventional gas systems is a worthwhile topic for further research.

AUTHIGENIC CEMENT

Stable isotopic analysis of calcite veins in cores of the Pottsville Formation has proven useful for characterizing the evolution of formation fluids and the physical and biotic processes that occurred therein (Pitman et al., 2003). The veins occur in natural fractures, specifically cleats in coal and joints in shale and sandstone. Carbon isotopic ratios (^{13}C) in calcite from coal-bearing strata and organic-rich shale provide information on pore fluid chemistry, specifically the dissolved inorganic carbon in formation water at the time of mineralization, as well as fractionation processes associated with late-stage methanogenesis (Gould and Smith, 1979 and Budai et al., 2002Pitman et al., 2003). Oxygen isotopic ratios (δ^{18}O), by comparison, can be used as paleothermometers, thus providing insight on when diagenesis occurred during the regional burial and unroofing history (Friedman and O'Neil, 1977 and Pitman et al., 2003).

Calcite in the cores tends to be enriched in ^{13}C, with δ^{13}C ratios ranging from about -5 to $+25$‰ (Fig. 16).Pitman et al. (2003) recognized that calcite is most enriched in ^{13}C adjacent to the recharge area along the southeastern basin margin and that veins in coal tend to be more enriched than those in shale and sandstone. They interpreted enrichment as the product of a biotically mediated fractionation process that favored enrichment of the formation fluid and calcite cement in ^{13}C during methanogenesis. A similar process appears to have occurred in the Atlantic Rim region of the greater Green River Basin, where formation water has been enriched in ^{13}C in concert with microbial CO_2 reduction in coal (McLaughlin et al., 2010). Thus, enrichment of the calcite in the fresh-water plumes is consistent with the augmentation of thermogenic gas with late-stage microbial gas. The low NO_3^- values coupled with elevated HCO_3^- in the fresh-water plumes (Fig. 7 and Fig. 8), moreover, may indicate nutrient depletion by intense microbial activity.

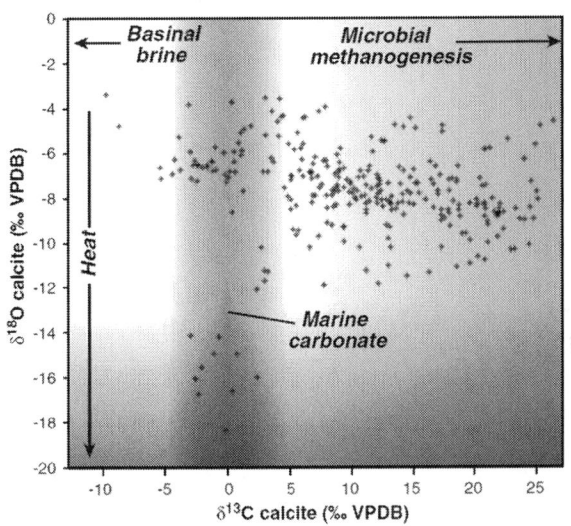

Figure 16: Cross-plot of $\delta^{18}O$ and $\delta^{13}C$ ratios in calcite vein fills from the Black Warrior CBM fields.

The bulk of the $\delta^{18}O$ data plot between − 12 and − 4‰, and no obvious correlation with $\delta^{13}C$ ratios is apparent (Fig. 16). A secondary cluster of data points plots with $\delta^{18}O$ values of − 14 to − 19‰, and all of these points have $\delta^{13}C$ ratios characteristic of normal marine carbonate. Plotting $\delta^{18}O$ values against depth reveal some basic relationships (Fig. 17). The calcite with $\delta^{18}O$ ratios of <− 14‰ show no relationship to modern burial depth. This calcite arguably formed in higher temperatures than the other calcite. One possibility is that it precipitated early in the regional unroofing history, perhaps before the reservoir strata reached the biogenic floor of the basin, which is typically where temperatures approach 80–100 °C (Carothers and Kharaka, 1978 and Shurr and Ridgley, 2002). The majority of the data have $\delta^{18}O$ ratios of >− 9‰, and maximum values appear to be depth- and hence temperature-limited.

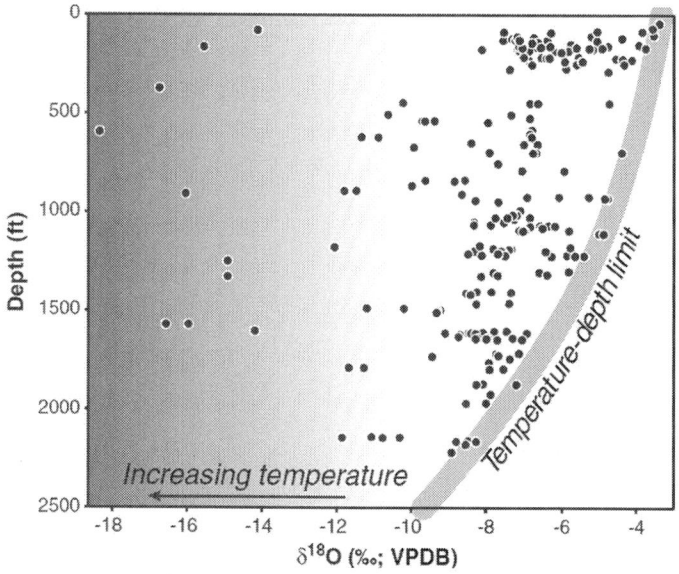

Figure 17: Plot of $\delta^{18}O$ ratios versus depth in calcite vein fills from the Black Warrior CBM fields.

The fractionation equations of Friedman and O'Neil (1977) and Hays and Grossman (1991) suggest that calcite with ^{18}O ratios of − 9 to − 4‰ precipitated at temperatures of 20 to 50 °C, which is compatible with modern reservoir temperatures in the Pottsville Formation (Pashin and McIntyre, 2003 and Pitman et al., 2003). Therefore, major calcite cementation in association with microbial CO_2 reduction apparently occurred late in the regional unroofing history as Pottsville strata approached modern burial depth. Analysis of the regional burial history suggests that calcite precipitation and methanogenesis may have begun during the Mesozoic (~ 150 Ma) and may continue today (Fig. 4).

DISCUSSION

Analysis of water and gas chemistry in the Pottsville Formation indicates a complex interrelationship among water composition, thermogenic hydrocarbon generation, and late-stage microbial methanogenesis. Throughout the study area, gas is produced at pipeline quality and so requires minimal processing prior to commercial distribution. The great variability of water chemistry, in contrast, necessitates that significant attention be paid to water processing and disposal, which must be conducted in an environmentally acceptable manner while controlling operational costs.

Chloride content is the dominant control on water quality in the Black Warrior CBM fields (Fig. 5 and Fig. 6) and is therefore a central concern for developing water management procedures. Most dissolved solids, as well as metallic and nonmetallic substances, correlate positively with TDS and chloride content. Thus, an effective water management strategy will naturally address these substances in concert with the chlorides. Importantly, the metals, nonmetals, and nitrogen compounds identified in the produced water are generally not of major concern from the standpoint of human health and safety (Table 2 and Table 3). Those that are (Pb, Se, and NH_3) have very low concentrations, save for a few isolated anomalies. An array of organic compounds occurs in the produced water (Table 4) and appears to be derived mainly from coal. A number of these compounds, including phenolic substances and PAH, are of environmental concern. Orem et al. (2007) pointed out that, at the low concentrations reported in produced water, the risk of acute exposure appears low. The risk of long-term chronic exposure is unclear, but the TDS content of water produced from the Black Warrior Basin precludes direct use for public water supply anyway.

Instream discharge to the Black Warrior River is currently the sole means of water disposal employed in the Black Warrior Basin CBM fields, and operators deploy significant facilities for the storage, processing, and disposal of produced water (Pashin, 2010a). A system of storage and treatment ponds is used to manage produced water. Synthetic membrane liners and monitoring wells ensure the integrity of impoundments. Pollutants are removed by suspension settling and aeration, which are crucial for mitigating environmental impact. Settling removes fine particles, such as clay and silt. Aeration enables oxidation and flocculation of metallic compounds and facilitates the evaporation, oxidation, and degradation of organic compounds in the produced water.

During the late 1980s, the membership of the Coalbed Methane Association of Alabama formed the Warrior Basin Environmental Cooperative, Incorporated. The cooperative operated an elaborate monitoring system that extended along 150 miles of the Black Warrior River. Toxicity testing (O'Neil et al., 1989 and O'Neil et al., 1993) was used to determine allowable chloride concentrations in the river. Consequently, discharge permits required that chloride concentrations in the river be < 230 mg/L. Concentrations were < 50 mg/L and never approached the regulatory limit. Accordingly, monitoring requirements were relaxed, and the cooperative was disbanded at the end of 1997.

Where TDS content and produced water volumes are high, water disposal concerns limit the ability to pump CBM wells to capacity (Pashin, 2010b). Reverse osmosis systems are used locally to process saline formation water, but the economic viability of these systems is currently being challenged by low natural gas prices. Artificial wetlands show promise for removing a broad range of contaminants from produced water, including chlorides, metals, and organic compounds (Rodgers and Castle, 2008 and Spacil et al., 2011), and Clemson University and Chevron are conducting a cooperative experimental program to determine the viability of artificial wetlands in the Black Warrior CBM fields. Considering the large volume of low TDS water produced from the CBM fields, artificial wetlands may enable processing of the water for beneficial use in industry and agriculture, as well as human consumption.

Fresh-water plumes are considered an essential part of the late-stage microbial methanogenic system in many CBM basins (Pashin, 2007, Rice, 1993, Scott et al., 1994 and Strapo´c et al., 2007). In the Black Warrior Basin, the plumes apparently provided vital pathways for anaerobic microbial consortia to colonize the interior of the basin. Within the basin, however, the evidence for late-stage biogenesis in hypersaline water suggests that the availability of nutrients and materials to sustain biologically mediated chemical reactions appears to be more important than salinity. A similar situation exists in the Illinois Basin, where microbial methanogenesis also has been documented in hypersaline water basinward of fresh-water plumes (Schlegel et al., 2011).

Significant hydrocarbon content in formation waters where coal lies in the oil generation window (Fig. 12) is one tangible source of nutrients that can sustain anaerobic microbes. Nitrate is another important nutrient that was identified in the formation water. Added to this is the possibility of thermally and microbially mediated denitrification reactions which, along with meteoric sources, can account for most if not all of the N_2 in the produced gas. Indeed, N_2 is the only significant impurity adversely affecting the calorific value of the product. Deionization of NH_4^+ may help complete the picture by providing a simple source of H_2 to augment the dominant organic sources in coal and formation water, thereby facilitating the reduction of CO_2 to CH_4.

SUMMARY AND CONCLUSIONS

CBM reservoirs of the Black Warrior Basin in Alabama are highly prolific, having produced more than 69×10^9 sm^3 of gas and $1.6 \times 10_6$ bbl of water since 1980. The coalbed gas industry in this area is dependent on in-stream disposal of the co-produced water. The chemistry of the produced water ranges from nearly potable sodium-bicarbonate water to hypersaline sodium-chloride water. Opportunities may exist for beneficial use of the nearly potable water. For example, technologies like artificial wetlands and reverse osmosis have potential to transform a significant part of the co-produced water into a resource that can be used for industrial and agricultural applications, as well as human consumption.

Integrated geological and geochemical analysis of produced water and gas reveals strong interrelationships among the regional geologic framework, water chemistry, and gas chemistry. These relationships must be understood to develop a robust strategy for produced water management. Water chemistry is influenced by a structurally controlled meteoric recharge area along the southeastern margin of the basin, and the most promising opportunities for beneficial use are near the recharge zone. Salinity increases with distance from the recharge zone, and the inability to pump wells to capacity where salinity is highest helps determine the economic viability of gas production.

Bulk chemistry of the produced gas is nearly invariant, whereas the stable isotopic composition of the gas varies regionally. The produced gas is enriched in ^{13}C in medium-volatile and low-volatile bituminous coal, suggesting a significant thermogenic component of gas generation. However, regional unroofing and cooling indicates that the coal would be greatly undersaturated with CH_4 without additional gas generated by late-stage microbial methanogenesis. The gas becomes depleted in ^{13}C as coal rank decreases to high volatile B bituminous away from the recharge area. Stable isotopic analysis of produced gas and calcite cement indicates that widespread late-stage microbial methanogenesis occurred primarily along a CO_2 reduction metabolic pathway.

Significant concentrations of organic compounds in the produced water appear to have helped sustain microbial communities. Ammonia and ammonium levels increase with TDS content and appear to have played a role in late-stage microbial methanogenesis and the generation of N_2. Indeed, a range of reactions involving organic matter, minerals, and formation water, may have contributed some N_2, which is the only significant nonhydrocarbon impurity in the produced gas. Biotic degradation of organic compounds may have been augmented by deionization of NH_4^+ thus providing multiple sources of H_2 for microbial CO_2 reduction.

ACKNOWLEDGMENTS

This research was funded by the National Energy Technology Laboratory of the U.S. Department of Energy under Award DE-FE0000888. The authors thank the coalbed methane producers and the Coalbed Methane Association of Alabama for their support, which included access to wells for sampling. Analytical work conducted at the USGS in Reston, VA was supported by the USGS Energy Resources Program (Brenda Pierce, Program Manager). We thank Anne Bates (USGS) for acetate analyses. Jennifer McIntosh and an anonymous reviewer provided many helpful comments and suggestions that greatly improved the quality of this contribution. Any use of trade, firm, or product names is for descriptive purposes only and does not imply endorsement by the U.S. Government. This paper is Boone Pickens School of Geology Contribution 2013-5.

REFERENCES

1. Ayers Jr., W.B., Kaiser, W.A. (Eds.), 1994. Coalbed methane in the Upper Cretaceous Fruitland Formation, San Juan Basin, Colorado and New Mexico. Texas Bureau of Economic Geology Report of Investigations, 218 (216 pp.).

2. Barker, D.S., 1964. Ammonium in alkali feldspars. Am. Mineral. 49, 851–854.

3. Bates, B.L., McIntosh, J.C., Lohse, K.A., Brooks, P.D., 2011. Influence of groundwater flowpaths, residence times, and nutrients on the extend of microbial methanogenesis in coal beds: Powder River Basin, USA. Chem. Geol. 284, 45–61.

4. Budai, J.M., Martini, A.M., Walter, L.M., Ku, T.C.W., 2002. Fracture-fill calcite as a record of microbial methanogenesis and fluid migration: a case study from the Devonian Antrim Shale, Michigan Basin. Geofluids 2, 163–183.

5. Bustin, A.M.M., Bustin, R.M., 2008. Coal reservoir saturation: impact of temperature and pressure. AAPG Bull. 92, 77–86.

6. Carothers, W.W., Kharaka, Y.K., 1978. Aliphatic acid anions in oil-field waters— implications for the origin of natural gas. AAPG Bull. 62, 2441–2453.

7. Carroll, R.E., Pashin, J.C., Kugler, R.L., 1995. Burial History and Source-Rock Characteristics of Upper Devonian Through

Pennsylvanian Strata, Black Warrior Basin, Alabama. Alabama Geological Survey Circular, 187 (29 pp.).

8. Debajyoti, P., Skrzypek, G., 2007. Assessment of carbonate-phosphoric acid analytical technique performed using GasBench II in continuous flow isotope ratio mass spectrometry. Int. J. Mass Spectrom. 262, 180–186.

9. Duit, W., Jansen, J.B.H., van Breemen, A., Bos, A., 1986. Ammonium micas in metamorphic rocks as exemplified by Dome de L'Agout (France). Am. J. Sci. 286, 702–732.

10. Elder, C.H., Deul, M., 1974. Degasification of the Mary Lee coalbed near Oak Grove, Jefferson County. Alabama by vertical borehole in advance of mining. U.S. Bureau of Mines Report of Investigations, 7968 (21 pp.).

11. Ellard, J.S., Roark, R.P., Ayers Jr., W.B., 1992. Geologic Controls on Coalbed Methane Production: An Example From the Pottsville Formation (Pennsylvanian Age), Black Warrior Basin, Alabama, U.S.A. In: Beamish, B.B., Gamson, P.D. (Eds.), Symposium on coalbed methane research and development in Australia, 1. James Cook University, Townsville, Australia, pp. 45–61.

12. Eugster, H.P., Munoz, J., 1966. Ammonium micas: possible sources of atmospheric ammonia and nitrogen. Science 151, 683–686.

13. Flores, R.M., Rice, C.A., Stricker, G.D., Warden, A., Ellis, M.S., 2008. Methanogenic pathways of coal-bed gas in the Powder River Basin, United States: the geologic factor. Int.

14. J. Coal Geol. v. 76, 52–75.

15. Friedman, I., O'Neil, J.R., 1977. Compilation of Stable Isotope Fractionation Factors of Geochemical Interest, In: Fleischer, M. (Ed.), Data of Geochemistry, 6th ed. U.S. Geological Survey Professional Paper, 440-KK (12 pp.).

16. Galloway, J.N., Dentener, F.J., Capone, D.G., Boyer, E.W., Howarth, R.W., Seitzinger, S.P., Asner, G.P., Cleveland, C.C., Green, P.A., Holland, E.A., Karl, D.M., Michaels, A.F., Porter,

J.H., Townsend, A.R., Vörösmarty, C.J., 2004. Nitrogen cycles: past, present, and future. Biogeochemistry 70, 153–226.

17. Gould, K.W., Smith, J.W., 1979. The genesis and isotopic composition of carbonates associated with some Permian Australian coals. Chem. Geol. 24, 137–150.

18. Graves, S.L., Patton, A.F., Beavers, W.M., 1983. Multiple zone coal degasification potential in the Warrior coal field of Alabama. Gulf Coast Association of Geological Societies

19. Transactions, 33, pp. 275–280.

20. Hall, F.E., Zhou, C., Gasem, K.A.M., Robinson, R.L., Yee, D., 1994. Adsorption of Pure Methane, Nitrogen, and Carbon Dioxide and Their Binary Mixtures on Wet Fruitland Coal. Society of Petroleum Engineers, paper SPE 29194, pp. 329–344.

21. Hays, P.D., Grossman, E.L., 1991. Oxygen isotopes in meteoric calcite cements as indicators of continental paleoclimate. Geology 19, 441–444.

22. Hunt, J.M., 1979. Petroleum Geochemistry and Geology. Freeman, San Francisco (617 pp.). Jones, E.P., Voytek, M.A., Corum, M.D., Orem, W.H., 2010. Stimulation of methane generation from nonproductive coal by addition of nutrients or a microbial consortium. Appl. Environ. Microbiol. 76, 7013–7022.

23. Jüntgen, H., Karweil, J., 1966. Gasbildung und gasspeicherung in Steinkohlenflözen, teilen 1 und 2. Erdol und Kohle-Erdgas-Petrochemie 19, 339–344.

24. Jüntgen, H., Klein, J., 1975. Entstehung von erdgas aus kohligen sedimenten. Erdöl und Kohle, Erdgas Petrochemistrie, Ergängsband 1, 52–69.

25. Juster, T.C., Brown, P.E., Bailey, S.W., 1987. NH3-bearing illite in very low-grade metamorphic rocks associated with coal, northeastern Pennsylvania. Am. Mineral. 72, 555–565.

26. Kaiser, W.R., Hamilton, D.S., Scott, A.R., Tyler, R., 1994. Geological and hydrological controls on the producibility of coalbed methane. J. Geol. Soc. Lond. 151, 417–420.

27. Karl, D.M., Michaels, A.F., 2001. Nitrogen Cycle. In: Steele, J.H., Turekian, K.K., Thorpe, S. (Eds.), Encyclopedia of Ocean Sciences, 4. Academic Press, New York, pp. 1876–1884.

28. Levine, J.R., Telle, W.R., 1989. A Coalbed Methane Resource Evaluation in Southern Tuscaloosa County, Alabama. School of Mines and Energy Development Research Report, 89-1. University of Alabama, Tuscaloosa, Alabama (90 pp.).

29. Martini, A.M., Walter, L.M., Ku, T.C.W., Budai, J.M., McIntosh, J.C., Schoell, M., 2003.

30. Microbial production and modification of gases in sedimentary basins: a geochemical case study from a Devonian shale gas play, Michigan basin. AAPG Bull. 87, 1355–1375.

31. Martini, A.M., Walter, L.M., McIntosh, J.C., 2008. Identification of microbial and thermogenic gas components from Upper Devonian black shale cores, Illinois and Michigan basins. AAPG Bull. 92, 327–339.

32. McFall, K.S., Wicks, D.E., Kuuskraa, V.A., 1986. A Geological Assessment of Natural Gas From Coal Seams in the Warrior Basin, Alabama. Gas Research Inst., Topical Rept. GRI 86/0272 (80 pp.).

33. McKee, C.R., Bumb, A.C., Koenig, R.A., 1988. Stress-Dependent Permeability and Porosity of Coal and Other Geologic Formations. SPE Formation Evaluation. 81–91 (March 1988).

34. McLaughlin, J.F., Frost, C.D., Sharma, S.S., 2010. Geochemical analysis of Atlantic Rim water, Carbon County, Wyoming: New applications for characterizing coalbed natural gas reservoirs. AAPG Bull. 95, 191–217.

35. Mingram, B., Hoth, P., Lüders, V., Harlov, D., 2005. The significance of fixed ammonium in Palaeozoic sediments for the generation of nitrogen-rich natural gases in the North

36. German Basin. Int. J. Earth Sci. 94, 1010–1022.

37. Oh, M.S., Taylor, R.W., Coburn, T.T., Crawford, R.W., 1988. Ammonia evolution during oil shale pyrolysis. Energy Fuels 2, 100–105.

38. O'Neil, P.E., Harris, S.C., Drottar, K.R., Mount, D.R., Fillo, J.P., Mettee, M.F., 1989. Biomonitoring of a produced water discharge from the Cedar Cove degasification field, Alabama. Alabama Geological Survey Bulletin, 135 (195 pp.).

39. O'Neil, P.E., Harris, S.C., Mettee, M.F., Shepard, T.E., McGregor, S.W., 1993. Surface discharge of wastewaters from the production of methane from coal seams in Alabama: the Cedar Cove model. Alabama Geological Survey Bulletin, 155 (259 pp.).

40. Orem, W.H., Tatu, C.A., Lerch, H.E., Rice, C.A., Bartos, T.T., Bates, A.L., Tewalt, S., Corum, M.D., 2007. Organic compounds in produced waters from coalbed natural gas wells in the Powder River Basin, Wyoming, USA. Appl. Geochem. 22, 2240–2256.

41. Pashin, J.C., 2007. Hydrodynamics of coalbed methane reservoirs in the Black Warrior Basin: key to understanding reservoir performance and environmental issues. Appl. Geochem. 22, 2257–2272.

42. Pashin, J.C., 2008. Coal as a Petroleum Source Rock and Reservoir Rock. In: Ruiz, I.S., Crelling, J.C. (Eds.), Applied Coal Petrology—The Role of Petrology in Coal Utilization. Elsevier, Amsterdam, pp. 227–262.

43. Pashin, J.C., 2010a. Mature Coalbed Natural Gas Reservoirs and Operations in the Black Warrior Basin of Alabama. In: Reddy, K.J. (Ed.), Coalbed Natural Gas: Energy and Environment. Nova Science Publishers, Haupage New York, pp. 31–57.

44. Pashin, J.C., 2010b. Variable gas saturation in coalbed methane reservoirs of the Black Warrior Basin: implications for exploration and production. Int. J. Coal Geol. 82, 135–146.

45. Pashin, J.C., Groshong Jr., R.H., 1998. Structural control of coalbed methane production in Alabama. Int. J. Coal Geol. 38, 89–113.

46. Pashin, J.C., McIntyre, M.R., 2003. Temperature–pressure conditions in coalbed methane reservoirs of the Black Warrior Basin, Alabama, U.S.A: implications for carbon sequestration and enhanced coalbed methane recovery. Int. J. Coal Geol. 54, 67–183.

47. Pashin, J.C., Ward II, W.E., Winston, R.B., Chandler, R.V., Bolin, D.E., Richter, K.E., Osborne, W.E., Sarnecki, J.C., 1991. Regional Analysis of the Black Creek-Cobb Coalbed-Methane Target Interval, Black Warrior Basin, Alabama. Alabama Geological Survey Bulletin, 145 (127 pp.).

48. Pashin, J.C., Carroll, R.E., Hatch, J.R., Goldhaber, M.B., 1999. Mechanical and Thermal Control of Cleating and Shearing in Coal: Examples From the Alabama Coalbed Methane Fields, USA. In: Mastalerz, M., Glikson, M., Golding, S. (Eds.), Coalbed Methane: Scientific, Environmental and Economic Evaluation. Kluwer Academic Publishers, Dordrecht, Netherlands, pp. 305–327.

49. Pashin, J.C., Jin, Guohai, Payton, J.W., 2004. Three-Dimensional Computer Models of Natural and Induced Fractures in Coalbed Methane Reservoirs of the Black Warrior Basin. Alabama Geological Survey Bulletin, 174 (62 pp.).

50. Pitman, J.K., Pashin, J.C., Hatch, J.R., Goldhaber, M.B., 2003. Origin of minerals in joint and cleat systems of the Pottsville Formation, Black Warrior Basin, Alabama: implications for coalbed methane generation and production. AAPG Bull. 87, 713–731.

51. Reddy, K.J., 2010. Coalbed Natural Gas. Energy and Environment. Nova Science Publishers, Haupage, New York (511 p.).

52. Rice, D.D., 1993. Composition and Origins of Coal-Bed Gas. In: Law, B.E., Rice, D.D. (Eds.), Hydrocarbons from coal. AAPG Studies in Geology, 38, pp. 159–184.

53. Rice, C.A., 2003. Production waters associated with the Ferron coalbed methane fields, central Utah: chemical and isotopic composition and volumes. Int. J. Coal Geol. 56, 141–169.

54. Rice, C.A., Ellis, M.S., Bullock Jr., J.H., 2000. Water Co-Produced With Coalbed Methane in the Powder River Basin. Wyoming: preliminary compositional data. U.S. Geological

55. Survey Open File-Report 00-372.

56. Rodgers Jr., J.H., Castle, J.W., 2008. Constructed wetland treatment systems for efficient and effective treatment of contaminated waters for reuse. Environ. Geosci. 15, 1–8.

57. Schlegel, M.E., McIntosh, J.C., Bates, B.L., Kirk, M.F., Martini, A.M., 2011. Comparison of fluid geochemistry and microbiology of multiple organic-rich reservoirs in the Illinois Basin, USA: evidence for controls on methanogenesis and microbial transport. Geochim. Cosmochim. Acta 75, 1903–1919.

58. Scott, A.R., 1993. Composition and Origin of Coalbed Gases From Selected Basins in the United States. 2004 International Coalbed Methane Symposium Proceedings.

59. University of Alabama, Tuscaloosa, pp. 207–222.

60. Scott, A.R., 2002. Hydrogeologic factors affecting gas content distribution in coal beds. Int. J. Coal Geol. 50, 363–387.

61. Scott, A.R., Kaiser, W.R., Ayers Jr., W.B., 1994. Thermogenic and secondary biogenic gases, San Juan Basin, Colorado and New Mexico—implications for coalbed gas producibility. AAPG Bull. 78, 1186–1209.

62. Seidle, J., 2011. Fundamentals of Coalbed Methane Reservoir Engineering. Pennwell Corporation, Tulsa (470 pp.).

63. Shurr, G.W., Ridgley, J.L., 2002. Unconventional biogenic gas systems. AAPG Bull. 86, 1939–1969.

64. Spacil, M.M., Rodgers Jr., J.H., Castle, J.W., Chao, W.Y., 2011. Performance of a pilot-scale constructed wetland treatment system for selenium, arsenic, and low molecular weight organics in simulated fresh produced water. Environ. Geosci. 18, 145–156.

65. Strapo´c, D., Mastalerz, M., Eble, C., Schimmelmann, A., 2007. Characterization of the origin of coalbed gases in

southeastern Illinois Basin by compound-specific carbon and hydrogen stable isotope ratios. Org. Geochem. 38, 267–287.

66. Taylor, G.H., Teichmüller, M., Davis, A., Diessel, C.F.K., Littke, R., Robert, P., 1998. Organic Petrology. Gebrüder Borntraeger, Berlin (704 pp.).

67. Telle, W.R., Thompson, D.A., Lottman, L.K., Malone, P.G., 1987. Preliminary Burial-Thermal History Investigations of the Black Warrior Basin: Implications for Coalbed Methane and Conventional Hydrocarbon Development. 1987 Coalbed Methane Symposium Proceedings. University of Alabama, Tuscaloosa, pp. 37–50.

68. Thamdrup, B., Dalsgaard, T., 2002. Production of N2 through anaerobic ammonium oxidation coupled to nitrate reduction in marine sediments. Appl. Environ. Microbiol. 68, 1312–1318.

69. Thomas, W.A., 1985. The Appalachian–Ouachita connection: Paleozoic orogenic belt at the southern margin of North America. Annu. Rev. Earth Planet. Sci. 13, 175–199.

70. Thomas, W.A., 1988. The Black Warrior Basin. In: Sloss, L.L. (Ed.), Sedimentary cover—North American craton. Geological Society of America, The Geology of North, America, D-2, pp. 471–492.

71. Thomas, W.A., 1995. Diachronous thrust loading and fault partitioning of the Black Warrior foreland basin within the Alabama recess of the Late Paleozoic Appalachian-Ouachita thrust belt. SEPM Special Publication, 52, pp. 111–126.

72. Thomas, W.A., Kanda, R.V.S., O'Hara, K.D., Surles, D.M., 2008. Thermal footprint of an eroded thrust sheet in the southern Appalachian thrust belt, Alabama, USA. Geosphere 4, 814–828.

73. Van de Graaf, A.A., Mulder, A., De Bruijn, P., Jetten, M.S.M., Robertson, L.A., Kuenen, J.G., 1995. Anaerobic oxidation of ammonium is a biologically mediated process. Appl. Environ. Microbiol. 61, 1246–1251.

74. Van Voast, W.A., 2003. Geochemical signature of formation waters associated with coalbed methane. AAPG Bull. 87, 667–676.

75. Vinson, D.S., McIntosh, J.C., Ritter, D.J., Blair, N.E., Martini, A.M., 2012. Carbon isotope modeling of methanic coal biodegradation: metabolic pathways, mass balance, and the role of sulfate reduction, Powder River Basin, USA. Geol. Soc. Am. Abstr. Programs 44 (7), 466.

76. Whiticar, M.J., 1996. Stable isotope geochemistry of coals, humic kerogens and related natural gases. Int. J. Coal Geol. 32, 191–215.

77. Whiticar, M.J., 1999. Carbon and hydrogen isotope systematic of bacterial formation and oxidation of methane. Chem. Geol. 161, 291–314.

78. Whiticar, M.J., Faber, E., Schoell, M., 1986. Biogenic methane formation in marine and freshwater environments: CO_2 reduction vs. acetate fermentation—isotopic evidence. Geochim. Cosmochim. Acta 50, 693–703.

79. Winston, R.B., 1990a. Vitrinite Reflectance of Alabama's Bituminous Coal. Alabama Geological Survey Circular, 139 (54 pp.).

80. Winston, R.B., 1990b. Preliminary Report on Coal Quality Trends in Upper Pottsville Formation Coal Groups and Their Relationships to Coal Resource Development, Coalbed Methane Occurrence, and Geologic History in the Warrior Coal Basin, Alabama. Alabama Geological Survey Circular, 152 (53 pp.).

Partial Coal Pyrolysis and its Implication to Enhance Coalbed Methane Recovery, Part I: An Experimental Investigation

Yidong Cai[a], Dameng Liu[a], Yanbin Yao[a], Zhentao Li[a], and Zhejun Pan[b]

[a]Coal Reservoir Laboratory of National Engineering Research Center of CBM Development & Utilization, China University of Geosciences, Beijing 100083, China

[b]CSIRO Earth Science and Resource Engineering, Private Bag 10, Clayton South, Victoria 3169, Australia

ABSTRACT

This paper examines the feasibility of combining a process known as enhanced methane recovery with partial coal pyrolysis to improve the petrophysics of coal seams and ultimately extract higher methane yields with accompanying pyrolysis gases. Partial pyrolysis for coal gas generation changes the pore and fracture structure, which in turn affect the permeability. A series of laboratory experiments on three coal rank samples monitored the changes in pore structure and permeability accompanying coal pyrolysis. Thermogravimetry–mass spectrometry (TG–MS) analysis evaluated mass loss and product composition. The pore and fracture structure evolution was determined by a combination of mercury intrusion porosimetry (MIP), scanning electron microscope (SEM) and methane adsorption capacity measurements on heat-treated coal blocks of ~100 g. The pore volume and methane adsorption capacity of LRC specimen (0.56% $R_{o,m}$) with 10 °C/min and a hold time of 30 min experienced slight changes during the heating process from 25 °C to 400 °C, but when heated from 400 °C to 800 °C, the pore volume in the LRC specimen greatly increased and the mercury-determined total porosity went from 36% at 400 °C to 43% at 800 °C. The permeability of the specimens at the temperature range of 300–400 °C increased exponentially with temperature due to the generated pore–fracture system. The sample LRC (800 °C) with the highest mercury-determined pore volume possessed the lowest methane capacity (19.45 cm³/g) due to the maximum adsorption volume of pyrolyzed coal obtained from the Langmuir model was related not only to the pore structure but also to the extent of graphitization. Therefore, they may have significant implications for enhanced coalbed methane (CBM) recovery.

INTRODUCTION

Coalbed methane recovery before coal extraction is very important from greenhouse gas emission, assistance in development of gas industry, safety and economic of mining point of view [1], [2] and [3]. Enhanced coalbed methane (ECBM) recovery from coal seams may consist of a multi-branch horizontal well, with applications of CO_2 sequestration, hydrofracturing, biotechnology and heating (steam injection and coal burning) etc. to store gases or enhance permeability/methane production. These methods have associated environmental and economic challenges. Coal pyrolysis is an alternative approach and is a component of underground coal gasification (UCG) [4], [5] and [6]. This method of gas production produces CH_4, H_2 and CO_2 from the thermochemical decomposition and gasification of organics at high temperatures in the absence of oxygen, which is similar as kerogen pyrolysis in petroleum field [7], [8],[9] and [10]. A similar approach where only the pyrolysis stage is performed could be a promising option for enhancing CBM recovery. Different physical and thermochemical transformations occur at elevated temperatures: (1) the first stage (25–300 °C), is the dry gas phase. Here, moisture and a small amount of adsorbed gases (including CH_4 and N_2, etc.) desorb from the matrix pores and cleat the system until ~200 °C. Thus, the coal structure does not change significantly. From 200 °C to 300 °C, thermal decomposition occurs for the low-rank coals such as lignite; (2) the second stage (300–550 °C), the coal pyrolysis phase has decomposition reactions, forming gases and tars. Before 400 °C, coal may soften, and viscous plastic mass can form. In the range from 400 to 550 °C, coal gas evolves and coal tar precipitates from the thermally decomposed products. The residual substance in coal gradually stiffens and solidifies as char. Gas produced in 450–550 °C range contains light aromatic hydrocarbon and long chain fatty mass. The petrophysics of coals (including pore structure, fractures and permeability etc.) changes during this phase [6], [11], [12] and [13]. The condensation reaction during the process of the formation of char is not very obvious; (3) in the third stage (550–1000 °C), polycondensation reactions occur (also called secondary generation phase), as well as the carbocoal transition phase. During this phase, carbocoal aromatic sizes increase, the arrangement of aromatic layers often become more ordered, true density increases and a high degree of aromatization occurs along with an of increasing metallic luster. The gas generation here is associated with the thermal decomposition of coal molecule functional groups. Complex organics are pyrolyzed into liquids and gases, which flow/burst out of the particles. Additionally, this stress and mass loss creates more pores and fractures. Thus, the permeability of the pyrolyzed coal is changed.

To accurately and continuously acquire the suitable maximum temperature for enhancing CBM recovery, thermogravimetry coupled with a mass spectrometry (TG–MS) was adopted to analyze the organic decomposition products from different rank coals at elevated temperatures (25–1200 °C) to obtain structural information. Secondly, methane adsorption profiles were determined by a programmed gas adsorption analysis rig. At the same time, the pore structures including pore volume, pore size distribution, porosity and permeability of coals during pyrolysis are investigated with increased temperatures from 25 °C to 800 °C. The feasibility of partial coal pyrolysis for low rank coal (LRC) will be focused due to its low gas content and shallow burial depth.

EXPERIMENTAL METHODS

Coal Sampling

Three coal samples with a volume of approximately 40 × 40 × 40 cm^3 were directly collected from three different rank coals including one low-rank coal (LRC, 0.56% $R_{o,m}$), one medium rank coal (MRC, 1.68% $R_{o,m}$) and one high rank coal (HRC, 2.59% $R_{o,m}$). LRC originated from the southern Junggar basin, NW China; MRC is collected from Ordos basin, North China; and HRC comes from Qinshui basin, North China. All samples were carefully packed and taken to the laboratory for experiments. Each sample was carefully preserved in an intact form.

TG–MS during Coal Pyrolysis

Thermogravimetry–mass spectrometry (TG–MS) analysis was conducted by a Rigaku TG–DTA coupled with an Omnistar MS. TG was linked to the programmed furnace, which can collect the continuous weights of samples and execute the data analysis. The TG has a temperature range of 25–1200 °C. The MS has a Nier type enclosed ion source, two detectors and a triple mass filter, which was controlled by computer. The two rigs were coupled by a transfer line leading from the TG to the MS. Generated gases in the TG rig can flow into the MS rig through the transfer line in a few seconds later. The MS need just very small fractions of the gases [14]. The TG conditions used to study the relationship between the structure and gas compositions were heating rate of 10 °C/min at the temperatures of 25–1200 °C; sweep nitrogen, 60 cm³/min; retention time, 30 mins; constant sample volume weighing 25–50 mg. The MS was scanned over a range of 0–100 amu with measurement intervals of approximately 19 s. The characteristics of the products of coal pyrolysis were determined vs. the elevated temperature in the multiple ion detection modes.

Microscopy and MIP Analysis

Vitrinite reflectance ($R_{o,m\%}$) and fracture analyses were conducted as previous research [15]. Coal compositions, proximate and ultimate analyses of used coals are documented in Table 1. The pore morphology was acquired using scanning electron microscope (SEM) and pore structure analysis for different rank coals (LRC, MRC and HRC) at different temperatures (25 °C, 200 °C, 400 °C, 600 °C and 800 °C) was performed with a PoreMasterGT60 which automatically registers pressure, pore radius and mercury injection volume. The SEM pictures of LRC sample was selected to be a representative due to the same trend for all three rank coals. Previous research [16] and [17] found that coal compressibility has an effect on mercury intrusion porosimetry (MIP) results especially when pressure is greater than 20 MPa. Data processing and the coal compressibility correction procedures are the same as in our previous work[18] and [19]. Assuming that pores are composed of a variety of cylindrical pores, the relationship between the pore radius and pressure can be acquired by the Washburn Equation [20]. On the basis that cumulative mercury injection volume, pore radius and space distribution can be inferred from the measured mercury injection curves.

Table 1: Vitrinite reflectance, proximate and ultimate analysis of different rank coals

Samples No.	$R_{o,m}$ (%)	Proximate analysis wt.% (ad)				Ultimate analysis wt.% (daf)			
		M	V	A	FC	C	H	N	St
LRC	0.56	0.74	9.58	12.71	76.97	74.80	4.49	3.07	1.08
MRC	1.68	1.19	11.64	6.22	80.95	81.92	4.40	4.03	0.23
HRC	2.59	2.24	5.34	3.50	88.92	90.96	2.76	1.67	0.13

Note: M-Moistures; V-Volatiles; A-Ash; FC-Fixed Carbon; C-Carbon; H-Hydrogen; N-Nitrogen; St-Total sulfur; ad-air dry basis; daf-dry ash for free basis.

Methane Adsorption Measurement

Methane adsorption on coal samples was conducted at 25 °C with the pressure range from 0 MPa to 10 MPa. Before the adsorption experiment, the crushed sample (60–80 mesh) was dried overnight at 50 °C and then degassed at 25 °C for 1 h under a vacuum. Approximately 100 g were used for each measurement. Methane adsorption isotherms were measured using the Isotherm Measurement System (KT100-40HT). The experimental procedure and the helium calibration for pore volume in coals were described in the previous research [15] and [21]. Adsorbed methane at a given pressure was similar to the measurement of the pore volume, which was finished when the pressure approached 10 MPa. Each process was repeated two or three times for data accuracy [22] and [23] and was analyzed using the Langmuir model to obtain adsorption parameters for each sample [24].

RESULTS AND DISCUSSION

Microscopy and SEM

The vitrinite reflectance values ($R_{o,m}$%) for the coal samples are shown in Table 1. They range from thermally immature 0.56–2.59% in the southern Junngar basin, southeastern Ordos basin and southern Qinshui basin. Fig. 1 shows SEM micrographs of coal samples at elevated temperatures. There was a slight difference in the pore morphology at the different temperatures (25 °C, 200 °C, 400 °C, 600 °C and 800 °C). Compared to LRC (25 °C), LRC (200 °C) and LRC (400 °C), the surface morphology of LRC (600 °C) and LRC (800 °C) looked spongy, indicating that the porosity of pyrolyzed coal was developed and the development of pores depended on the heating temperatures, which was consistent with the result of mercury intrusion porosimetry. The heat treatment during pyrolysis leads to changes in pore structure, such as the pores enlargement and pore creation. The closure or partial closure of pores due to the thermal expansion at low temperature (<200 °C) may also occur [13]. As shown in Fig. 1, the pore morphology determined by SEM reaches the plateau of its maximum size at 600 °C, which means that 600 °C is enough to improve the coal reservoir for LRC sample. Thus it would provide additional gas adsorption or storage sites and affect the gas transport in an important way.

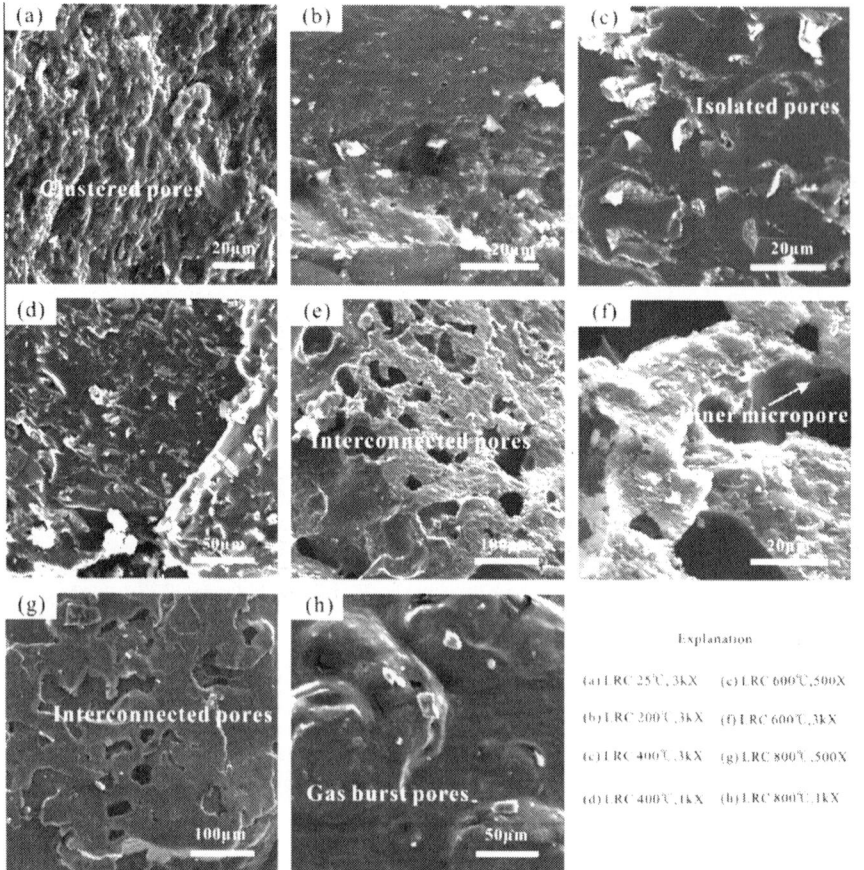

Figure 1: Pore morphology of LRC sample at different temperatures from SEM.

Gas Generation and Gas Types

During the heating process, many changes occurred including coal structure, coal organic composition, volatiles and moistures adsorbed in the coal, which may physically alter the movement of methane in coals during the adsorption process [25] and [26]. Thus Fig. 2 depicts the weight change of the original coal due to the change of thermal reaction during the process of coal pyrolysis at a temperatures range of 25–1200 °C under the flow of nitrogen. The heating not only removed volatiles and moistures but also resulted in alterations of the pore structure during the low heating process (lower than 200 °C), suggesting that the removal of volatiles and moistures produced more micropores. From TG–MS curve, the weight change was negligible during this low heating stage.

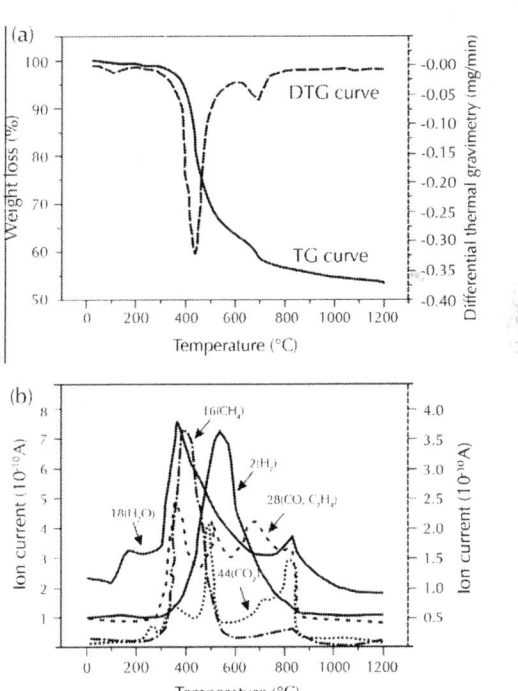

Figure 2: Weight change and gas composition during pyrolysis for LRC sample.

Previous research [6] and [13] found that dehydration mainly occurs below 350 °C, together with a small amount of other gases. Extensive evolution of gases had taken place in the range of 400–800 °C, which caused the stacking in coal structure [27]. The major gases generated during coal pyrolysis contain H_2O, CH_4, CO_x, C_nH_m, H_2 (Fig. 2). Oxocarbon decreased gradually with increasing temperature, reaching up to 70% when the temperature hit 350 °C, and then going down to 10–30% when the temperature arrived at 500 °C. In the pyrolysis process, C2 (C_nH_m, $n = 2$) was reduced when the temperature reached 450 °C. The methane emissions increased with increasing temperature from 300 °C, reaching a peak at ~400 °C and accounting for 50–60% of the mass. The generation of CH_4 originated from the thermal reaction when the temperature is normally lower than 700 °C:

$$Coal\text{-}CH_3 + H^* \rightarrow CH_4 \tag{1}$$

$$C(solid) + 2H_2 \rightarrow CH_4 \tag{2}$$

Methane was mainly generated at this stage, as shown in Fig. 2 and Fig. 3. Thus a suitable treatment temperature for enhanced coalbed methane recovery should be <700 °C, which was also accompanied by the changes in accessibility of the pore volume. Although it could be expected that heating could improve the pore accessibility to some extent and generate much gases, the accompanying tar production may condense and could block the flow paths of the pores [27], [28] and [29], thus the well completion should start after the temperature of coal reservoir has cooled down.

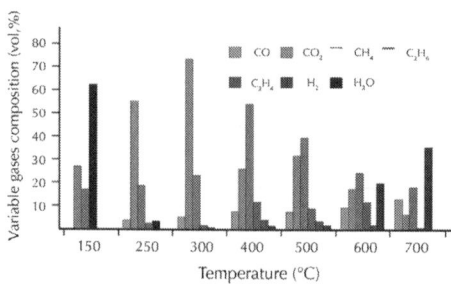

Figure 3: The variable gases composition from the pyrolysis of LRC sample.

Petrophysics of Coal Reservoir during Pyrolysis

Coal as a porous medium with economic interest has been widely researched for many years, especially for its petrophysics including pores, fractures, diffusivity and permeability [11], [15], [26] and [30]. However, the petrophysical changes associated with coal pyrolysis are much different from those occurring in normal stress situations during mining or CBM exploitation [12], [31] and [32]. Complex physical and chemical changes happen during the process of coal pyrolysis, and these greatly alter the fracture and pore structures of the coal along with the mass generation including gases or heavy hydrocarbons [12] and [33].

Pore Structure and Fractures

Pore size distribution with elevated temperatures is shown in Fig. 4, which was acquired by MIP. To provide a better understanding of the effects of pore structure on both gas adsorption and permeability, a combined classification for coal pore size is used in this work: super micropores (<2 nm), micropores (2–10 nm), mesopores (10–10^2 nm), macropores (10^2–10^3 nm), super macropores (10^3–10^4 nm) and microfractures (>10^4 nm) as previous research [19]. Pore volume and average pore radius with increased temperatures are documented in Table 2. The temperatures of 400 °C and 600 °C create significant changes in the pore structure. When the temperature was lower than 200 °C, few thermal reactions took place. Due to the remove of partial moistures and volatiles, slightly more super micropores and micropores volume was generated. However, pore volume, average pore radius and porosity increased markedly when the pyrolysis temperature reached 400 °C and 600 °C. Between these two temperatures, a large increase was shown in the pore radius and porosity, which means that a strong dilation of the fractures and pores volume took place.

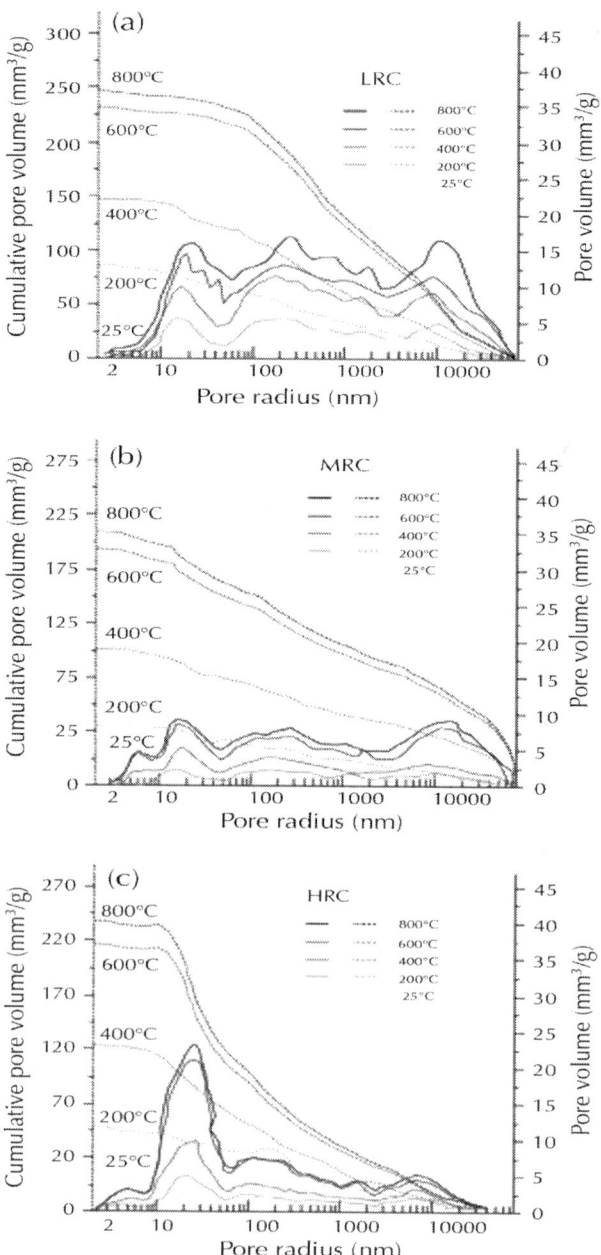

Figure 4: Pore size distribution of MIP for different rank coals at different temperatures.

Table 2: Pore structure and Langmuir parameters of equilibrium isotherms ($T = 25$ °C) of LRC

Sample No.	Sample density (g/cm³)	Pore volume (mm³/g)	Average pore radius (mm)	Langmuir model		
				R^2	V_L (cm³/g)	P_L (MPa)
LRC (25 °C)	1.39	39.2	45.3	0.995	14.13	1.69
LRC(200 °C)	1.41	85.7	50.6	0.992	13.56	1.79
LRC (400 °C)	1.47	140.5	90.8	0.993	17.81	1.07
LRC(600 °C)	1.53	230.4	110.6	0.996	19.45	0.79
LRC (800 °C)	1.57	246.8	151.7	0.991	11.93	2.06

Pore size distribution with temperature can be divided into three stages. The first is the low temperature stage from 25 °C to 200 °C where the total pore volume is relatively low, and the original micro and macro pores are predominant, occupying ~70% of the total. The second stage ranges from 200 °C to 400 °C where all pores increase with increasing temperature. However, the macropore volume increases faster than that of other pore sizes. The macropores content increased from ~20% to 50% from the low to medium temperature stages, which should effectively improve the permeability of specimen. In the last high temperature stage from 400 °C to 600 °C, the micropores volume linearly increased with temperature and they comprised ~35% of the total volume. Although the pores volume for other scale pores also increased with increasing temperatures, the amplitude was relatively less than the micropores. Previous research[34] proved that for low temperature pyrolysis of Yallourn brown coal that the micropore size distribution was not significantly changed, while for high temperature pyrolysis a remarkable increase of the microporosity was observed due to the release of inorganic gases after the tar formation stage.

Fractures are very important to coal permeability, especially for the process of CBM exploitation and coal mining [35], [36], [37] and [38]. For LRC sample, fractures are less developed, have low density and were characterized by isolated, orthogonal or Y-shaped structures, but with relatively good connectivity and infrequent mineralization. The fractures can be divided into four types based on our previous classification[15] and [39]: Type A, with width (W)5 μm and length (L) 10 mm; Type B, with W 5 μm and L 10 mm; Type C, with W 5 μm and L 300 μm, and Type D, with W 5 μm and L 300 μm. The fractures evolution along with elevated temperatures is shown in Fig. 5, which shows that the main contribution to the increase of total fractures should be type A and type D. Previous research reveals that temperature can effectively improve the fracture area even when it does not exceed 100 °C [40]. There is a rapid increase for fractures at 300 °C and then the fracture frequency is relatively stable until 800 °C, which means the suitable temperature for enhanced CBM should be in range of 300–800 °C, especially from the perspective of permeability.

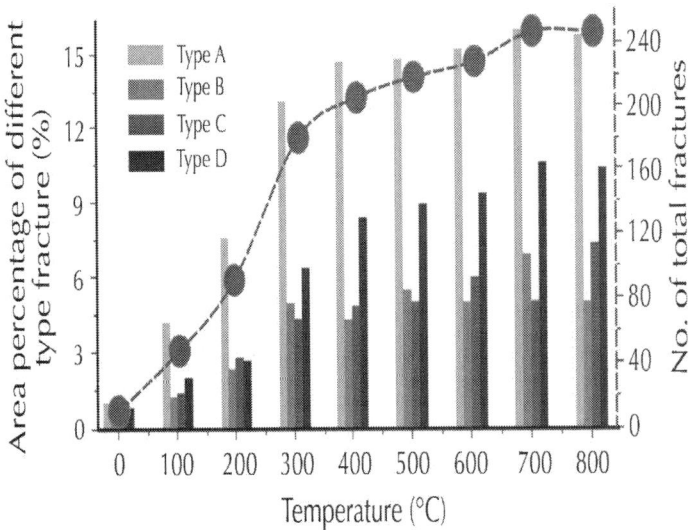

Figure 5: Area of different type fractures and total fractures with temperatures for LRC sample.

Porosity and Permeability

Previous research [41] conducted porosity measurements on two subbituminous coals during pyrolysis and obtained porosities of 32.46% at 400 °C, 43.37% at 600 °C and 59.89% at 800 °C, respectively. One Chinese steam coal with the porosity of 22.89% at 400 °C and 36.69% at 600 °C [12] shows that temperature has the same effect on porosity for different coals, which are also analogous to our measurements. The coal porosity is made up of pores of different sizes over a broad range, up to five orders of magnitude (Fig. 4). The pore size distribution and total porosity of different rank coals at elevated temperatures are illustrated in Figs. 4 and 6a, which show that the transition point of porosity during pyrolysis increases with the increasing coal rank. The porosity of the LRC and MRC samples decreases slightly as temperatures rise from 25 °C to 100 °C, which may be due to thermal expansion. The pyrolysis does not play a role in this period, as the temperature is still far lower than the critical temperature [13]. It then increases quickly from

100 °C to 400 °C due to the remove of partial moistures, volatiles, initial pyrolysis and the significant generation of pores and fractures. Both physical mechanisms and pyrolysis play important roles in this great increase of porosity. The important temperature range for coal pyrolysis is from 400 °C to 800 °C, during which abundant gas and coal tar are released because of the depolymerization and decomposition. This becomes the period for the highest mass loss (Fig. 2). The temperature for transition point of porosity increases following the increasing coal rank, which means that the coals went through strong hypozonal metamorphism need more energy to depolymerize and decompose organics. Although the escape of the gas or coal tar will produce more pores and fractures and thus cause an increase of its porosity, the sedimentation of the coal tar or other products will occupy some pores and fissures after 700 °C. Thus the maximum temperature for enhancing CBM recovery by partial coal pyrolysis is limited to 700 °C, especially for low-rank coal.

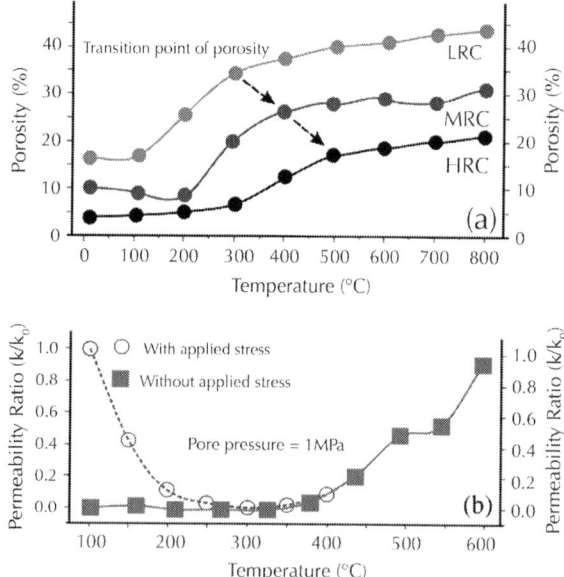

Figure 6: Porosity and permeability ratio change with increased temperatures (data of square points [12] and data of round points [43]) during coal pyrolysis.

Given the significance of coal permeability, many researchers focus on the permeability of coal and obtained valuable results from both experiments and theoretical modeling [19], [35], [36] and [38]. Previous research [42] found an empirical relationship between the meso-pores and the permeability on the basis of the pore parameter and permeability in coal during pyrolysis. The relationship between the pores and the permeability without applied stress was evidenced by previous research [43] and phenomenologically divided into three stages: slight fluctuation stage (25–300 °C); exponential increase stage (300–400 °C) and linear increase stage (400–600 °C). The relationship between the pores and the permeability with applied stress was evidenced by previous research [42] and [43] and phenomenologically divided into two stages (Fig. 6b): exponential decrease stage (25–300 °C) and exponential increased stage (300–400 °C). These two experiments show that stress has a strong impact on coal permeability at the low temperature stage (<300 °C), there is no significant result for high temperatures due to the experimental constraints. To make the permeability under different conditions (with and without applied stress) comparable, the raw permeability data was transferred to be the permeability ratio. These two experiments also show that there is an exponential trend for permeability after 300 °C, which means the pyrolysis has a larger effect on permeability than thermal expansion at the high temperature range, up to 400 °C.

Methane Adsorption

When the coal specimens were heated to 200 °C, some pores such as interconnected pores and pores accessible to gases were easily jammed by the accumulation of volatiles [23] and thermal expansion. Based on the TG–MS experiments, the volatiles should be partially removed before 200 °C, which may effectively decrease the porosity of super micropores and micropores. Thus it could reduce the methane adsorption capacity to 13.56 cm^3/g. With the temperature further increased to 400 °C, the moistures and volatiles within the coal matrix escaped and partial organics were pyrolyzed, which effectively increased the pore volume and improved the methane adsorption capacity up to 17.81 cm^3/g. Although the LRC (800 °C) has a larger pore volume than other samples at lower temperatures (25 °C, 200 °C, 400 °C and 600 °C), the sample LRC (800 °C) had a lower methane adsorption capacity than others due to the graphitization. Thus the pore structure and the extent of graphitization worked together for the methane adsorption capacity.

The methane adsorption at 25 °C in the pressure range of 0–10 MPa, shown in Fig. 7, was collected by the volumetric method. The uptake of methane drastically increased at relatively low pressures (<1 MPa), but then increased slowly with the pressure further increasing, which was comparable to previous results [15] and [25]. The Langmuir model fitted the data well and the parameters obtained from the Langmuir model are summarized in Table 2, which have been widely used in evaluating the methane adsorption capacities of coal reservoirs. Langmuir volume (V_L) is the maximum adsorption volume of the coal and Langmuir pressure (P_L) is the pressure when the adsorption volume reaches 50% of the maximum adsorption volume. For the LRC sample at different temperatures, the sample LRC (800 °C) showed the lowest methane adsorption capacity of 11.93 cm^3/g, and the sample LRC (800 °C) with the highest pore volume did not possess the highest methane capacity of 19.45 cm^3/g. The maximum adsorption volume obtained from the Langmuir model was not only related to the pore structure but also to the extent of graphitization.

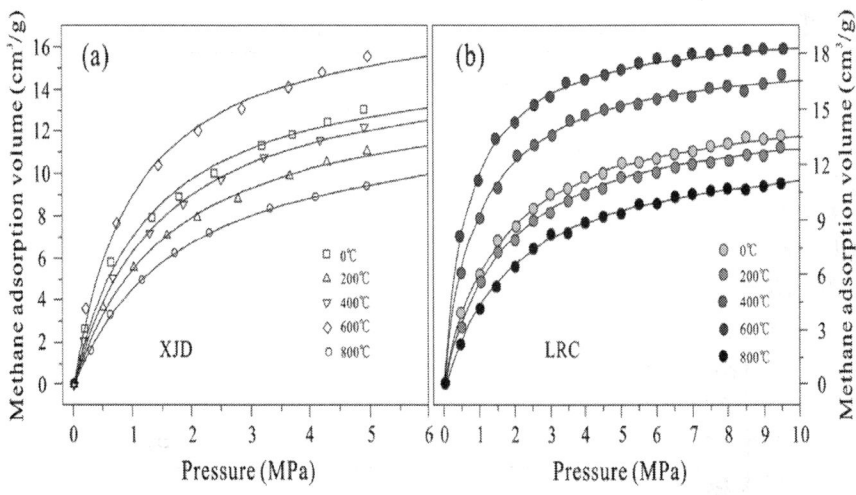

Figure 7: Comparison of methane adsorption profile obtained by applying the Langmuir model at 25 °C; (a) data of XJD sample [23] and (b) data of LRC sample.

Implication to Enhanced Coalbed Methane Recovery

From the experiments, we gain knowledge about the petrophysics development and the pore structure alteration, which results in the variation of methane adsorption capacity and changes in permeability. The roles that pore structure and graphitization play in the methane adsorption of coals under elevated temperatures will lead to a better understanding of the methane adsorption mechanisms of the coals. In addition, such experiments help to establish qualitative and semi-quantitative relationships between the weight loss and the amount of generated gases (Fig. 8), thus providing significant information for methane production from coal seams. Assuming the coal volume is a constant (C), the new generated gas content can be calculated by:

$$GC_{added} = \frac{v - v'}{m - m'} - \frac{v}{m}$$

(3)

where GC_{added} is the new generated gas content (cm³/g); v is the original gas volume (cm³); v' is the new generated gas volume (cm³); m is the original coal weight (g); m' is the lost weight during pyrolysis (g).

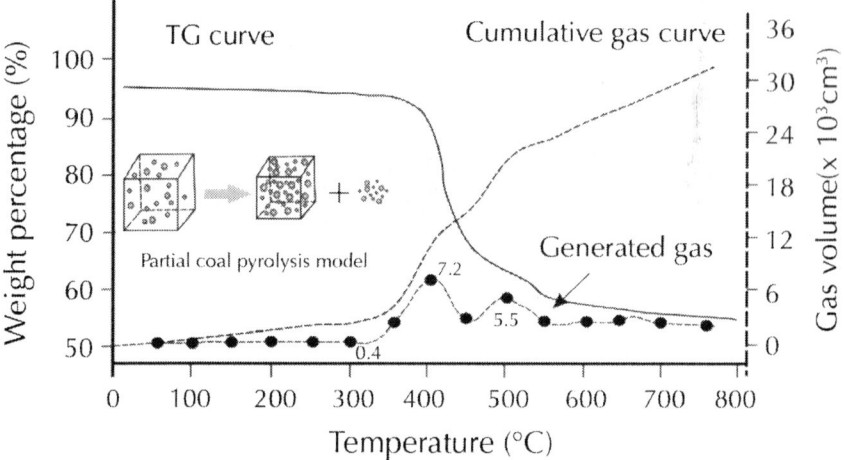

Figure 8: Partial coal pyrolysis model and new generated gas content model for LRC.

The valuable gases are extracted from the pyrolysis of coals with increasing temperature. Therefore, the crucial factor of the partial pyrolysis is the petrophysics of the coal reservoir and the migration of the produced gases. The major advantage of partial pyrolysis for enhanced CBM recovery is that organics in coal are converted into easily handled gases and become a form of clean energy [4] and [5]. Coal pyrolysis will occur at a critical temperature; the products including the tars (liquid phases) and gases generated during the process will then permeate into the production wells (Fig. 9).

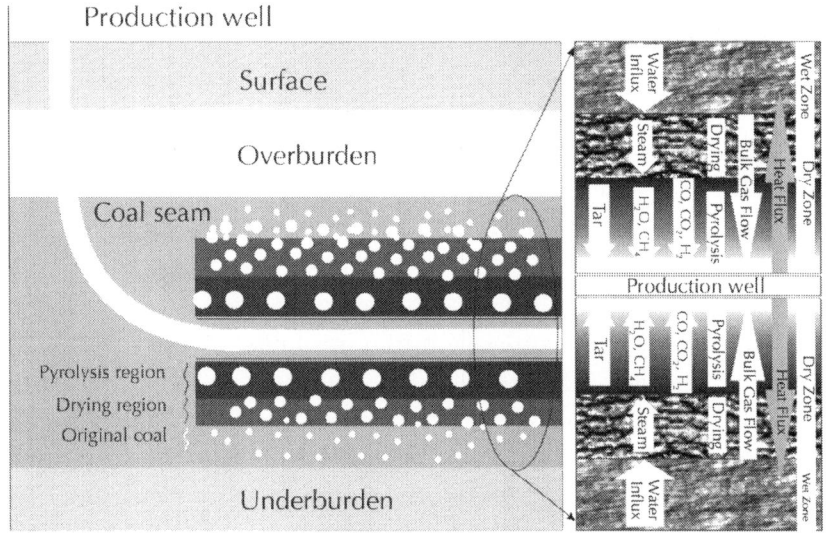

Figure 9: Thermal wave propagation during in-situ partial coal pyrolysis for enhanced coalbed methane recovery which demonstrates the different regions.

CONCLUSIONS

Pore structure, fracture features, permeability and methane adsorption capacity of coal during partial pyrolysis process were investigated to provide the prospective results for enhancing CBM recovery. Multiple methods (including SEM, TG–MS, MIP etc.) were used to study the petrophysical changes accompanying partial pyrolysis. The TG–MS revealed that the main temperature range for CH_4 generation is from 300 °C to 600 °C. Combing the suitable maximum temperature for improving petrophysics of coal reservoir, the most favorable maximum temperature for enhancing CBM recovery should be <700 °C. The pore structure significantly changed and the pore volume was greatly developed with elevated temperatures, especially after 400 °C. Methane adsorption results in partial pyrolysis were well described by the Langmuir model, which demonstrated that the high adsorption capacity of methane was related not only to the pore structure but also to the extent of graphitization. However, many mechanisms of partial coal pyrolysis are still unresolved; more work is therefore needed to provide a better understanding of this phenomenon.

ACKNOWLEDGEMENTS

This research was funded by the National Major Research Program for Science and Technology of China(Grant nos. 2011ZX05034-001 and 2011ZX05062-006), the United Foundation from the National Natural Science Foundation of China and the Petrochemical Foundation of PetroChina (Grant no. U1262104), the Program for New Century Excellent Talents in University (Grant no. NCET-11-0721), the Foundation for the Author of National Excellent Doctoral Dissertation of PR China (Grant no. 201253), the Fundamental Research Funds for Central Universities (Grant no. 2652013006) and the Research Program for Excellent Doctoral Dissertation Supervisor of Beijing (Grant no. YB20101141501).

REFERENCES

1.   Karacan CÖ, Ulery JP, Goodman GVR. A numerical evaluation on the effects of impermeable faults on degasification efficiency and methane emissions during underground coal mining. Int J Coal Geol 2008;75:195–203.

2.   Karacan CÖ, Ruiz FA, Cote M, Phipps S. Coal mine methane: a review of capture and utilization practices with benefits to mining safety and to greenhouse gas reduction. Int J Coal Geol 2011;86:121–56.

3.   Xu H, Tang DZ, Tang SH, Zhao JL, Meng YJ, Tao S. A dynamic prediction model for gas–water effective permeability based on coalbed methane production data. Int J Coal Geol 2014;121:44–52.

4.   Li CZ. Some recent advances in the understanding of the pyrolysis and gasification behaviour of Victorian brown coal. Fuel 2007;86:1664–83.

5.   Stanczyk K, Howaniec N, Smolinski J, Kapusta K, Grabowski J, Rogut J. Gasification of lignite and hard coal with air and oxygen enriched air in a pilot scale ex situ reactor for underground gasification. Fuel 2011;90:1953–62.

6.   Bhutto AW, Bazmi AA, Zahedi G. Underground coal gasification: from fundamentals to applications. Prog Energ Combust 2013;39(1):189–214.

7.   Tiwaria P, Deo M, Lin CL, Miller JD. Characterization of oil shale pore structure before and after pyrolysis by using X-ray micro CT. Fuel 2013;107:547–54.

8.   Eglinton TI, Larter SR, Boon JJ. Characterisation of kerogens, coals and asphaltenes by quantitative pyrolysis—mass spectrometry. J Anal Appl Pyrol 1991;20:25–45.

9.   Durand-Souron C, Boulet R, Durand B. Formation of methane and hydrocarbons by pyrolysis of immature kerogens. Geochim Cosmochim Ac 1982;46(7):1193–202.

10.   Marshall CP, Kannangara GSK, Wilson MA, Guerbois JP, Hartung-Kagi B, Hart G. Potential of thermogravimetric analysis coupled with mass spectrometry for the evaluation of kerogen in source rocks. Chem Geol 2002;184(3–4):185–94.

11.   Tarasevich YI. Porous structure and adsorption properties of natural porous coal. Colloid Surface A 2001;176(2–3):267–72.

12.   Zhao YS, Qu F, Wan ZJ, Zhang Y, Liang WG, Meng QR. Experimental investigation on correlation between permeability variation and pore structure during coal pyrolysis. Transp Porous Med 2010;82:401–12.

13.   Yu YM, Liang WG, Hu YQ, Meng QR. Study of micro-pores development in lean coal with temperature. Int J Rock Mech Min Sci 2012;51:91–6.

14. Rubel AM, Jagtoyen M, Stencel JM, Ahmed SN, Derbyshire FJ. TG-MS for characterization of activated carbons from coal. In: Symposium on analytical techniques for characterizing coal and coal conversion products American Chemical Society Washington DC, Mefling, August 23–28, 1992. pp. 1206–13.

15. Cai YD, Liu DM, Yao YB, Li JQ, Qiu YK. Geological controls on prediction of coalbed methane of No. 3 coal seam in Southern Qinshui Basin, North China. Int J Coal Geol 2011;88:101–12.

16. Li Y, Lu G, Rudolph V. Compressibility and fractal dimension of fine coal particles in relation to pore structure characterisation using mercury porosimetry. Part Part Syst Charact 1999;16:25–31.

17. Comisky JT, Santiago M, McCollom B, Buddhala A, Newsham KE. Sample size effects on the application of mercury injection capillary pressure for determining the storage capacity of tight gas and oil shales. CSUG/SPE 2011;149432:1–23.

18. Cai YD, Liu DM, Yao YB, Li JQ, Liu JL. Fractal characteristics of coal pores based on classic geometry and thermodynamics models. Acta Geol Sin-Engl 2011;85:1150–62.

19. Cai YD, Liu DM, Pan ZJ, Yao YB, Li JQ, Qiu YK. Pore structure and its impact on CH4 adsorption capacity and flow capability of bituminous and subbituminous coals from Northeast China. Fuel 2013;103:258–68.

20. Washburn EW. The dynamics of capillary flow. Phys Rev 1921;17:273–83.

21. Luo JJ, Liu YF, Jiang CF, Chu W, Jie W, Xie HP. Experimental and modeling study of methane adsorption on activated carbon derived from anthracite. J Chem Eng Data 2011;56:4919–26.

22. Lee WH, Park JS, Sok JH, Reucroft PJ. Effects of pore structure and surface state on the adsorption properties of nano-porous carbon materials in low and high relative pressures. Appl Surf Sci 2005;246:77–81.

23. Feng YY, Jiang CF, Liu DJ, Chu W. Experimental investigations on microstructure and adsorption property of heat-treated coal chars. J Anal Appl Pyrol 2013;104:559–66.

24. Karacan CÖ. An effective method for resolving spatial distribution of adsorption kinetics in heterogeneous porous media: application for carbon dioxide sequestration in coal. Chem Eng Sci 2003;58:4681–93.

25. Pan ZJ, Connell LD, Camilleri M, Connelly L. Effects of matrix moisture on gas diffusion and flow in coal. Fuel 2010;89(11):3207–17.

26. Cai YD, Liu DM, Pan ZJ, Yao YB, Li JQ, Qiu YK. Petrophysical characterization of Chinese coal cores with heat treatment by nuclear magnetic resonance. Fuel 2013;108:292–302.

27. Arenillas A, Rubiera F, Pis JJ, Cuesta MJ, Iglesias MJ, Jiménez A, et al. Thermal behaviour during the pyrolysis of low rank perhydrous coals. J Anal Appl Pyrol 2003;68–69:371–85.

28. Bae JS, Bhatia SK, Rudolph V, Massarotto P. Pore accessibility of methane and carbon dioxide in coals. Energy Fuels 2009;23:3319–27.

29. Li C, Zhao J, Fang Y, Wang Y. Pressurized fast-pyrolysis characteristics of typical Chinese coals with different ranks. Energy Fuels 2009;23:5099–105.

30. Liu DM, Yao YB, Tang DZ, Tang SH, Che Y, Huang W. Coal reservoir characteristics and coalbed methane resource assessment in Huainan and Huaibei coal fields, Southern North China. Int J Coal Geol 2009;79:97–112.

31. Yin GZ, Jiang CB, Wang JG, Xu J. Combined effect of stress, pore pressure and temperature on methane permeability in anthracite coal: an experimental study. Transp Porous Med 2013;100:1–16.

32. Niu SW, Zhao YS, Hu YQ. Experimental ivestigation of the temperature and pore pressure effect on permeability of lignite under the in situ condition. Transp Porous Med 2014;101:137–48.

33. Charland JP, MacPhee JA, Giroux L, Pricea JT, Khanb MA. Application of TG-FTIR to the determination of oxygen content of coals. Fuel Process Technol 2003;81:211–21.

34. Matsuo Y, Hayashi J-I, Kusakabe K, Morooka S. In: Proceedings of the Eighth International Conference on Coal Science, Oviedo. Amsterdam: Elsevier; 1995.

35. Liu JS, Chen ZW, Elsworth D, Qu HY, Chen D. Interactions of multiple processes during CBM extraction: a critical review. Int J Coal Geol 2011;87(3–4):175–89.

36. Pan ZJ, Connell LD. Modelling permeability for coal reservoirs: a review of analytical models and testing data. Int J Coal Geol 2012;92:1–44.

37. Wang S, Elsworth D, Liu J. Permeability evolution during progressive deformation of intact coal and implications for instability in underground coal seams. Int J Rock Mech Min Sci 2013;58:34–45.

38. Cai YD, Liu DM, Mathews JP, Pan ZJ, Elsworth D, Yao YB, et al. Permeability evolution in fractured coal -Combining triaxial confinement with X-ray computed tomography, acoustic emission and ultrasonic techniques. Int J Coal Geol 2014;122:91–104.

39. Yao YB, Liu DM. Microscopic characteristics of microfractures in coals: an investigation into permeability of coal. Procedia Earth Planet Sci 2009;1(1):903–10.

40. Mathews JP, Pone JDN, Mitchell GD, Halleck P. High-resolution X-ray computed tomography observations of the thermal drying of lump-sized subbituminous coal. Fuel Process Technol 2011;92:58–64.

41. Tomeczek J, Gil S. Volatiles release and porosity evolution during high pressure coal pyrolysis. Fuel 2003;82(3):285–92.

42. Dana E, Skoczylas F. Gas relative permeability and pore structure of sandstones. Int J Rock Mech Min Sci 1999;36:613–25.

43. Li MM. The research on pyrolysis-penteation and microstructure of lignite. Master Thesis of Taiyuan University of Technology, 2012.

Chapter 7

Trends in Water Quality Variability for Coalbed Methane Produced Water

Katharine G. Dahm[a, b], Katie L. Guerra[a, b], Junko Munakata-Marr[a], and Jörg E. Drewes[a, c]

[a]Advanced Water Technology Center (AQWATEC), Civil and Environmental Engineering, Colorado School of Mines, Golden, CO 80401-1887, USA

[b]U.S. Bureau of Reclamation, Denver, CO 80225-0007, USA

[c]Technische Universität München, Chair of Urban Water Systems Engineering, 85748 Garching, Germany

ABSTRACT

Energy production from unconventional natural gas resources, such as coalbed methane, has the potential to generate significant water quantities for use in water-stressed areas to augment existing water supplies. Coalbed methane (CBM) produced water is generated from shallower formations than traditional oil and gas resources where water quality may be influenced by fresh water supplies in the area. Variability in produced water quality between wells and across geologic basins must be characterized in order to categorize water types appropriate for beneficial use. Principal component analysis (PCA) was applied to a composite geochemical database to identify indicators of variability in water composition and quality. Component analysis revealed that water quality indicators of variability were related to: (i) aquifer recharge that dilutes constituent concentrations (37%), (ii) dissolution of soluble aquifer minerals such as sodium and exchange of calcium and magnesium (13.8%), and (iii) coal depositional environment influence on chloride and trace metal fractions (14% of variability). Ternary relationships between $Na–Cl–HCO_3$ and $Na–Ca–Mg$ correlate to marine influence in the coal depositional environment and well proximity to recharge, respectively. Relationships identified in this study highlight water quality compositions with opportunities for beneficial use.

Graphical Abstract

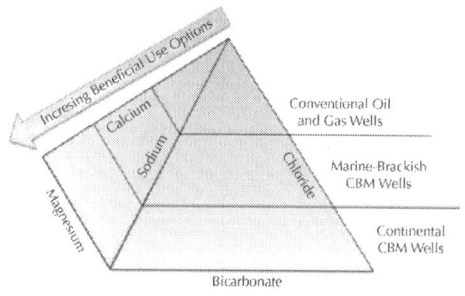

INTRODUCTION

In the western United States (US) where demand for fresh water resources exceeds current availability, frequent droughts coupled with the competing demand for municipal supplies and agricultural needs result in water shortages and potential conflict over existing water supplies (Bureau of Reclamation, 2003). Water generated during oil and gas production, termed produced water, is the largest waste stream from oil and gas production and over 80% of produced water is generated in the western US (Clark and Veil, 2009). Produced water is saline and usually disposed of through injection into the subsurface. Produced water is transported along oil and gas pipelines to central disposal locations for injection. In certain locations, produced water disposal is restricted due to regulations or available geologic formations. While coalbed methane (CBM) produced water often requires treatment to meet beneficial use requirements, its use can reduce the strain on conventional water sources when managed as part of a regional water portfolio. Beneficial use of produced water alleviates water shortages and increases disposal options for the oil and gas industry.

Areas such as the Central Valley in California are home to over 50,000 oil and gas wells that generate almost 55 billion gallons of water annually (Clark and Veil, 2009). Potential uses for produced water in these regions include on-site use for well drilling or hydraulic fracturing, livestock water, irrigation water, surface water augmentation, and drinking water applications. Based on the water quality information available from the United States Geological Survey (USGS) produced water ranges in TDS concentrations from 100 to 400,000 mg/L (United States Geological Survey, 2002). Over 71% of wells in the USGS database have TDS concentrations in excess of 50,000 mg/L, limiting both the types of applications for use and the desalination technologies capable of treatment.

CBM is an unconventional natural gas resource with large reservoirs in the water scare area of the Rocky Mountain region of the US. CBM wells are completed within coal seams and produce water to release hydrostatic pressure, allowing methane to desorb from the coal surface (Orem et al., 2007). Water is produced from within the coal seam itself due to the position of the screened interval of the well and the higher effective conductivity along coal cleats compared to the surrounding geologic formation (Rice, 2003). CBM basins commonly span large geographic areas and contain numerous coal seams, creating variable water chemistry among well fields. Coalbed methane produced water is generally less saline than conventional oil and gas produced waters because it is often produced from shallower formations that may interact with fresh water aquifers. Lower salinity and production larger volumes increase the potential for beneficial use of this energy industry byproduct.

Interaction with fresh water aquifers dilutes saline water associated with the coal formation at the time of deposition. Wells with active interaction with surface and groundwater supplies are more suitable for beneficial use. Wells with significant interaction with allocated surface and groundwater supplies are categorized as tributary in states where prior appropriations designate tributary waters to allocated downstream users through water rights, such as in the state of Colorado. In Colorado, where the withdrawal of coalbed methane produced water may lead to decreases in allocated supplies, regulations require the beneficial use of produced water to offset withdrawals. Researchers generally correlate tributary wells to produced water age, and a number of studies have focused on defining the age of coalbed methane produced water (Rice et al., 1989, Rice, 2003, Pashin, 2007, Rice et al., 2008, Cheung et al., 2009 and Kinnon et al., 2010). Wells with significant recharge offer better water quality for beneficial uses and increased sustainability of supply through recurring recharge connection.

Groundwater models predict a connection between allocated surface water resources and coal seam connectivity by estimating hydraulic conductivity and transmissivity through subsurface aquifers. Bethke and Johnson (2002) proposed a correction for the estimation of groundwater age related to the mixing of aquitard water and recharge flow through the subsurface. This estimate attributes groundwater age to mixing ratios instead of mixing rates to understand the relationship between groundwater flow and radiometric age. Coal seam water quality can be attributed to the water present at the time of formation (Van Voast, 2003). Therefore, the mixing of the formation water with fresh water sources gives an estimation of the groundwater age and interaction with fresh water sources. Using this theory, the suitability of produced water for beneficial use may be categorized based on indicators of interactions with fresh water resources.

A composite geochemical database containing 3255 CBM well entries, from the Atlantic Rim area of the Greater Green River Basin, the Powder River Basin, the Raton Basin, and the San Juan Basin, was created to characterize produced water quality to understand beneficial use potential in the Rocky Mountain region (Dahm et al., 2011). The database analysis identified constituents of concern for a number of beneficial use applications including trace metals, sodicity, and volatile compounds, such as BTEX. General trends by basin were identified and initial analysis indicated a link between geochemical origins of CBM water quality based on sub-surface groundwater interaction and beneficial use opportunities. Understanding of variations in water quality and correlations to geochemical factors to predict treatment requirements and end uses, however, were not explored.

Information in the database for 64 water quality parameters and constituents was sufficient to identify parameters that commonly exceed beneficial use requirements, such as total dissolved solids (TDS), temperature, iron, fluoride, trace metals and sodium adsorption ratio (SAR). Using principal component analysis (PCA) to determine the major sources of variability in CBM produced water, the following objectives were targeted to indicate well water quality suitability for beneficial use: (i) relate the principal components to environmental characteristics and basin attributes to determine identifiers for fresh water dilution, (ii) determine constituent relationships using commonly measured water quality parameters that describe water composition based on the principal components determined, and (iii) validate constituent relationships using publicly available data to determine whether trends identified in this study may be used to infer suitability of produced waters from other CBM basins for beneficial use.

MATERIALS AND METHODS

This study focused on three major CBM basins in the Rocky Mountain Region of the US: the Powder River Basin, the Raton Basin, and the San Juan Basin. The Atlantic Rim portion of the Greater Green River Basin was excluded from this study due to limited well entries, with only 34 well entries for the Greater Green River basin, constituting 1% of the total entries in the database. The first step in isolating and describing variable water compositions in CBM produced water was to use PCA to identify variability in the database. PCA describes the dataset variability by grouping variations into component coefficients, conducted using the Microsoft® Excel based add-on program StatistiXL. Correlation matrices were used because variables did not share consistent units. Eigenvalues greater than 1 were used to determine principal component scores and equations. Component coefficients were deemed significant if they exceeded 0.30 or were less than −0.30. Statistical boundaries were chosen to reflect similar studies of geochemical formation water data (Taulis, 2007).

The database variables used as inputs for the PCA included bicarbonate, calcium, chloride, iron, magnesium, pH, potassium, sodium, sulfate, total dissolved solids (TDS) and trace metals. The trace metals variable was constituted as the summation of the inorganic trace metal concentrations present in the database: aluminum, arsenic, cadmium, chromium, cobalt, copper, lead, lithium, manganese, molybdenum, nickel, selenium, silver, tin, titanium, vanadium, and zinc. Manganese, lithium, and aluminum exhibit the greatest concentrations among the seventeen trace metals, but each of the trace metal average database concentrations were less than 1 mg/L. Trace constituents, such as beryllium and antimony, were excluded from the trace metal variables if all well entries were non-detect. Non-detect or below detection limit data was adjusted to null to reflect the measurement method value of detection, for instance data less than the 0.01 mg/L detection limit were adjusted to zero during the statistical analysis, so non-detect trends could be observed. Other variables present in the original database but excluded from PCA due to limited number of well entries included: benzene, toluene, ethylbenzene, and xylene (BTEX), radionuclides, dissolved organic carbon (DOC), silica, and the anions bromide, fluoride and nitrate.

Ternary relationships focused on the use of commonly measured constituents and parameters, such as sodium, calcium, magnesium, chloride, bicarbonate, sulfate, and TDS, to describe the water composition trends determined to be related to each principal component. Based on literature review of each resulting system characteristic, ion combinations were identified that could be used to describe the chemical composition of each variable component. Ternary plots were created using multivariate fraction relationships from the geochemical composite database (Dahm et al., 2011). The OriginLab® OriginPro software was used to plot multivariate relationships between constituent fractions and display ternary relationships.

Finally, the constituent relationships created to describe water composition related to the principal components were validated using water quality information extracted from published studies on CBM produced water quality. Using the sources listed in Supplementary Data Table 1, a total of 139 sample analyses from CBM basins were selected to validate trends and describe aquifer connectivity. Data for sodium, calcium, magnesium, chloride, and bicarbonate were extracted from the peer-reviewed papers and plotted using the multivariate ternary relationships determined to describe the principal components.

RESULTS AND DISCUSSION

These results identify descriptive components that suggest aquifer connectivity based on water composition and were derived using

PCA. The results of the composite geochemical database analysis are described in the following sections.

Principal Component Analysis

PCA of CBM produced water quality for the Powder River, Raton and San Juan Basins isolated three principal components (PC-1, PC-2, and PC-3) that describe water quality variability. Coefficient scores for each component and eigenvalues are included in Supplementary Data Table 2. Using the eigenvalue cutoff of one, the variability in water quality described by the three components totaled 62%. Bar charts of the component scores include dotted lines to indicate the significance cutoff of ±0.30 (Fig. 1). Variables that exceeded these significance cutoff values are highlighted in color and labeled in each panel of the figure.

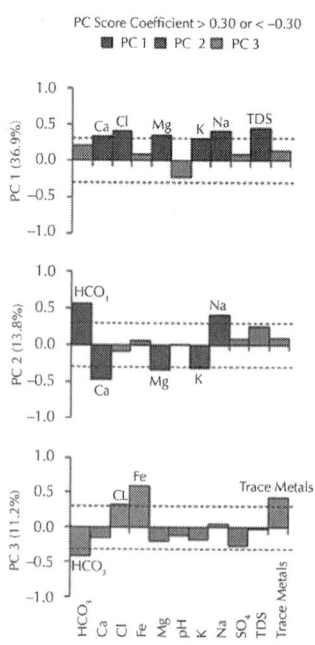

Figure 1: Score coefficients for the components of the PCA correlation analyses. The significance values ±0.3 are shown by dashed lines. Sig-

nificance variables are noted by labels above or below individual bars on each plot.

PC-1 describes 37% of the database variability. Based on the coefficient score cutoff of ±0.30, the correlation components vary positively with respect to calcium, chloride, magnesium, potassium, sodium and TDS. Bicarbonate, iron, pH, sulfate and trace metals are not considered significant for this component based on the ±0.3 cutoff. PC-2 describes 14% of the database variability. Based on the coefficient score cutoff, the correlation components vary positively with respect to bicarbonate and sodium and negatively with respect to calcium, magnesium and potassium. Finally, PC-3 describes 11% of the database variability. Based on the coefficient score cutoff, the correlation components vary positively with respect to iron, chloride and trace metals, and negatively with bicarbonate. Based on the PCA, sulfate and pH were the only variables that did not significantly influence any of the resulting principal components. TDS is considered a significant variable in PC-1, but is absent with respect to the other components. None of the variables act as significant influences to variability in all three components.

PCA Relationship to Environmental Characteristics

The principal components identified in the database PCA were assessed to determine the environmental attributes that contributed to each component of water quality variation. Fundamental environmental factors that were considered to impact water quality in CBM wells, included coal geochemistry (Bouška, 1981), geomicrobiology (Konhauser, 2007), and groundwater interactions (Garrels and Christ, 1965 and Freeze and Cherry, 1979). Previously published studies specific to CBM produced water chemistry were also used to classify environmental attributes of the system (Rice and Nuccio, 2000, Van Voast, 2003, Taulis, 2007, Johnston et al., 2008, Rice et al., 2008 and Cheung et al., 2009). These sources were analyzed to determine which environmental attributes described the resulting principal component trends. The principal components determined in section 3.1 provide information on key components and trends that describe database variability. The principal component variable relationships were assessed to determine how variable correlations could be explained by environmental attributes of the system. Environmental factors, such as the coal geochemistry, geomicrobiology, methane generation pathway, groundwater interactions, aquifer connectivity, fresh water recharge and depositional environment water quality, were assessed to determine the potential influence each characteristic had on produced water chemistry. Fig. 2 illustrates a number of environmental processes, such as fresh water recharge, aquifer interactions, and methane generation, and their potential impacts on CBM produced water chemistry. By matching available information regarding environmental attributes to the resulting principal component variable relationships, each principal component was attributed primarily to the following environmental processes.

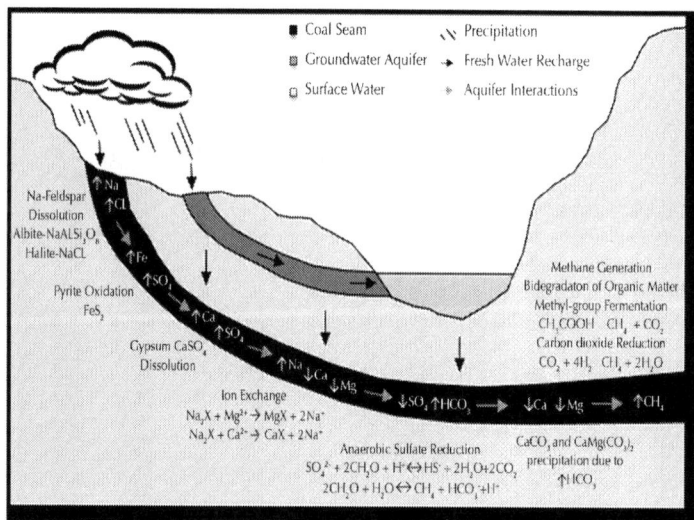

Figure 2: Aquifer interactions along the recharge flowpath and methane generation. Note: These processes are shown along the flowpath, but they do not necessarily occur sequentially.

Aquifer Recharge

Positive variations between almost all the variables suggest that the ion trends reflect an environmental characteristic that influences most constituents collectively. Based on the coal system characteristics examined, principal component 1 (PC-1) most likely describes the variability associated with dilution due to aquifer recharge. Coal fractures often create sufficient permeability to allow water and natural gas to flow through the formation, and coal seams can act as regional aquifers with specific recharge zones that convey groundwater through the coal seams. TDS concentrations are commonly lower near recharge zones where residence time is short, but increase with distance from the recharge zone and with depth below the surface (Johnston et al., 2008). Generally, aquifer recharge creates higher dissolved oxygen and lower TDS at close proximity to the location of recharge (Taulis, 2007 and Johnston et al., 2008) while higher TDS concentrations, bicarbonate, sodium and chloride concentrations are found as distance from recharge increases (Rice and Nuccio, 2000 and Van Voast, 2003).

Van Voast (2003) provided a summary of recharge pathways based on groundwater flow studies for the basins considered in this study. The Powder River basin has recharge along the eastern side of the basin, the Raton Basin exhibits a down gradient flow path toward the center of the basin, and the San Juan Basin experiences fresh water recharge in the Colorado part of the basin (Van Voast, 2003). Aquifer recharge is expected to cause dilution of most major ions. Component analysis in this study indicated aquifer recharge creates the largest variations in water quality between wells, coal seams and CBM basins.

Groundwater–mineral Interactions

Correlation components varying positively with respect to bicarbonate and sodium and negatively with respect to calcium, magnesium and potassium reflect a number of the aquifer mineral reactions indicated along the groundwater flow path in Fig. 2. For this reason the component score trends for principal component 2 (PC-2) are expected to be related to groundwater–mineral interactions. Groundwater interactions with mineral species vary along the groundwater flow path. For example, freshwater recharge can contact minerals, such as sodium and potassium feldspars, which dissolve in recharge waters. Freshwater recharge also can exchange ions with marine minerals, such as albite ($NaAlSi_3O_8$) and halite ($NaCl$), resulting in increased sodium concentrations while decreasing calcium and magnesium concentrations (Taulis, 2007). Bicarbonate and sulfate can enter the recharge flow system through carbonate and gypsum weathering, respectively. Microbial reduction of sulfate increases bicarbonate concentrations and thus increases the pH, decreasing the solubility of calcium and magnesium, causing precipitation (Van Voast, 2003). Water in coal seams becomes more sodic with depth due to the dissolution of sodium and the precipitation of calcium and magnesium (Larson and Daddow, 1984). Mineral-freshwater interactions along the groundwater flow path can be characterized by initial dissolution of calcium, magnesium, and sulfate, ion exchange of calcium and magnesium for sodium, microbial reduction of sulfate and increases in bicarbonate concentrations, which result in the precipitation of calcium, magnesium, barium and strontium (Van Voast, 2003 and Taulis, 2007). This shift in groundwater composition is evident in those basins where flow conditions have been determined, as concentrations nearer the recharge areas are generally lower for sodium and chloride and somewhat greater for calcium and magnesium than they are nearer the discharge areas (Van Voast, 2003). With groundwater–mineral interactions (PC-2) accounting for another 14% of variability in water quality, variability due to recharge and recharge interactions account for over 50% of the variability observed in the database.

Depositional Environment

Correlation components (principal component 3; PC-3) that vary positively with respect to iron, chloride and trace metals, and negatively with bicarbonate were difficult to match to the aquifer mineral interaction or the methane generation pathways described in Fig. 2. Further exploration into the presence of trace metals, chloride and iron indicated that this relationship was instead related to the coal depositional environment. CBM is produced in coal seams that were deposited in aquatic environments that range from continental freshwater sources to brackish or marine saline sources (Bouška, 1981). Formation water, present at the time of coal deposition, is contained in the coal seam pores. Continental or fresh water deposits are characteristically Na–HCO$_3$ type waters with higher calcium and magnesium concentrations and lower sodium and potassium concentrations relative to brackish or marine environments. Brackish and marine environments are generally Na–Cl type water with higher relative concentrations of sodium, chloride, boron and trace metals (Bouška, 1981, Van Voast, 2003 and Cheung et al., 2009). Marine environments also have higher relative concentrations of trace metals than continental environments (Cheung et al., 2009). Much of the bicarbonate in CBM systems is generated through sulfate reduction during coalification, resulting in generally high bicarbonate (greater than 500 mg/L) and low sulfate levels (less than 500 mg/L) characteristic of CBM produced water (Van Voast, 2003).

Continental depositional environments and biogenic methane formation pathways characterize the Powder River Basin (Rice et al., 2008). The produced water of this basin is mainly characterized as Na–HCO$_3$ type. The Raton Basin ranges in depositional environment from continental to brackish to marine (Van Voast, 2003) with predominantly thermogenic methane production (Cooper et al., 2007); its produced water exhibits elevated chloride concentrations with compositions ranging from Na–HCO$_3$ to Na–Cl. The San Juan Basin ranges in depositional environment from continental to brackish with a mix of biogenic and thermogenic methane production across the basin (Van Voast, 2003). The San Juan Basin exhibits a broader and more variable range of dominant salt characteristics than the other two basins, resulting in Na–HCO$_3$ to Na–HCO$_3$–Cl type waters (Dahm et al., 2011). The Raton and

San Juan basins also exhibit higher average iron concentrations than the Powder River.

Based on this analysis, environmental factors were identified to relate to the three principal components that account for 63% of the water quality variations. The geochemistry of the coal seam aquifer consists of a) solutes derived from the original depositional environment in which the coal was formed, b) solute dilution from freshwater recharge, and c) solutes derived from rock–mineral interactions along the flow paths of the recharge water. Over 95% of the solutes present in CBM produced water are sodium, bicarbonate and chloride (Dahm et al., 2011). Therefore, a majority of the constituents in solution are a result of the depositional environment, but variability in these constituents occurs as a result of recharge and aquifer mineral interactions.

Indicator Trends for Principal Components

Section 3.2 identified that the sources of water quality variability described by the PCA were related to recharge, aquifer mineral interactions, and the coal depositional environment. To determine which wells were influenced by these environmental characteristics, indicator plots were created based on the expected constituent relationships defined as part of each environmental attribute. These plots link recharge, aquifer mineral interactions, and the coal depositional environment to commonly measured water quality parameters.

Aquifer Recharge and Mineral Interactions Ternary Diagram (Na–Ca–Mg)

Aquifer recharge and groundwater–mineral interactions (PC-1 and PC-2) describe more than 50% of the variability observed in water chemistry in the CBM produced water database. Based on Fig. 2, the distance from recharge can be described by a decrease of calcium and magnesium and an increase of sodium due to interactions between minerals and the aquifer. Ternary diagrams were created to evaluate the relationship between sodium, calcium, and magnesium in the three basins (Fig. 3). Fig. 3 shows that calcium and magnesium influence is greater in the Powder River and San Juan basins than in the Raton Basin.

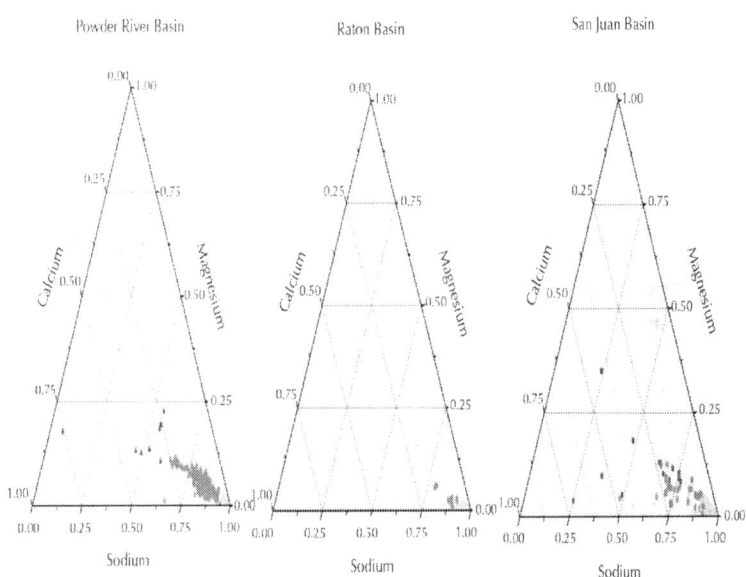

Figure 3: Ternary plots for the cations sodium, calcium, and magnesium for the Powder River Basin, the Raton Basin, and the San Juan Basin. The colors of the symbols on the graphs represent three ranges of SAR from the produced waters from the CBM wells. (For interpretation of the references to color in this figure legend, the reader is referred to the web version of this article).

The symbols on the ternary plots represent three different ranges for Sodium Adsorption Ratio (SAR). The SAR, used to describe sodicity, is an indicator parameter for the distribution between sodium, calcium and magnesium. SAR is calculated using Equation (1), where [Na⁺], [Ca²⁺], and [Mg²⁺] are in units of meq/L:

$$\text{Sodium Adsorption Ratio (SAR)} = [Na^+] \Big/ \sqrt{1/2x([Ca^{2+}] + [Mg^{2+}])} \tag{1}$$

Wells with high sodium to calcium and magnesium fractions exhibit higher SAR values. Wells exhibiting SAR less than 13 meet beneficial use requirements for irrigation, but most of the produced waters investigated exhibit ternary relationships strongly dominated by sodium or with SAR greater than 13 (Fig. 3). Well chemistries dominated by calcium and magnesium exhibit SAR values less than 1 and were rarely observed in any of the wells sampled, suggesting that the recharge pathways to the coal seams are not short enough to conserve shallow groundwater signatures.

Depositional Environment Ternary Diagram (Na–Cl–HCO₃)

Principal component (PC-3) is related to the effects of the depositional environment on variability. CBM produced water quality reflects formation waters regardless of formation lithology or age (Van Voast, 2003). CBM produced water is dominated by sodium, chloride and bicarbonate composing greater than 95% of the dissolved solids in all the basins studied. The depositional environment dictates the relationship between these three constituents. Therefore, the depositional environment directly affects the largest solute fractions in CBM produced water.

Ternary relationships between sodium, bicarbonate and chloride (Fig. 4) were used to differentiate major ion relationships between fresh water and marine-influenced basins. The distribution of data on the ternary diagram reflects the depositional environment that is characteristic of the basin. Continental basins dominated by Na–HCO_3 type waters plot along the bottom axis. Mixed fresh water and brackish water basins result in a span of salt fractions ranging from the bottom axis (Na–HCO_3 dominant) toward the apex (Na–Cl dominant). The continental or fresh water depositional environment results in Na–HCO_3 type salts and lower TDS in produced waters such as those observed in the Powder River Basin. The Raton Basin exhibits Na–HCO_3 to Na–Cl type waters with higher TDS due to brackish or marine influences on the produced water. The Raton Basin produced water demonstrates a distinct gradation in chloride and bicarbonate, which is well correlated with TDS concentrations. Higher TDS wells are Na–Cl type, while lower TDS wells are Na–HCO$_3$ type. The San Juan Basin produced water exhibits characteristics spanning a range of compositions due to a gradient of depositional environments ranging from continental to brackish or marine within this basin.

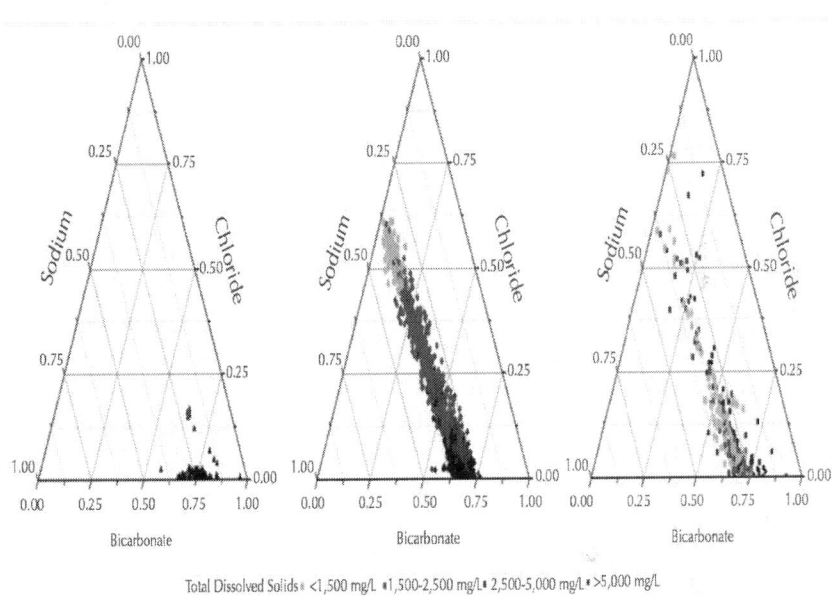

Figure 4: Ternary plots of sodium, chloride and bicarbonate in produced waters from the Powder River Basin, the Raton Basin, and the San Juan Basin. The colors of the symbols on the graphs represent four ranges of TDS from the produced waters from the CBM wells. (For interpretation of the references to color in this figure legend, the reader is referred to the web version of this article).

Validation of Indicator Trends using Additional Basins

Water quality data for 139 sample points from CBM basins world-wide were extracted from publicly available studies (Supplementary Data Table 1) and plotted using the same ternary relationships used to generate Fig. 3 and Fig. 4 (gray scale diagram points). These composite plots (Fig. 5 and Fig. 6) are described in this section. The database information from Dahm et al. (Dahm et al., 2011) has been combined and plotted on one diagram per relationship and set in black and gray scale to contrast against the additional points from the comparison data. Trends are described in general terms by increasing and decreasing influences of fresh water and marine water as illustrated in Fig. 6. In addition, 758 well points from conventional oil and gas wells from the USGS produced water database and 13 surface water samples from the Powder River Basin are also included for comparison (United States Geological Survey, 2002 and Jackson and Reddy, 2007).

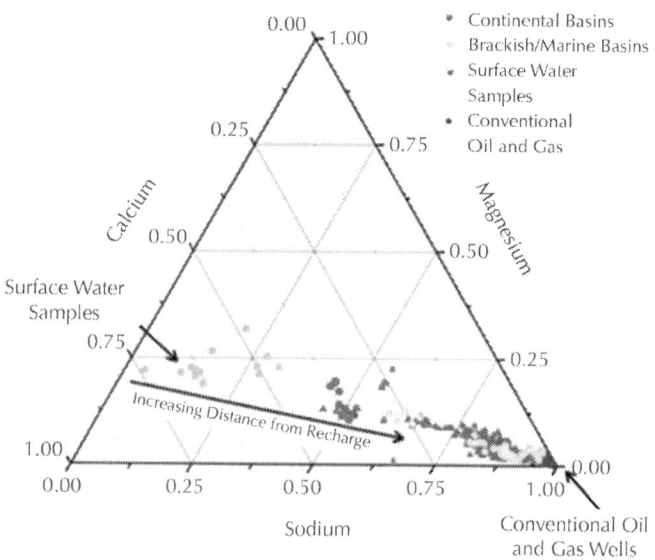

Figure 5: Ternary plots of sodium, magnesium and calcium for conventional oil and gas wells, surface waters from the Powder River Basin, and additional CBM produced water sources worldwide. Note; Grayscale points are CBM produced water from the database as previously shown in Fig. 4 of this study.

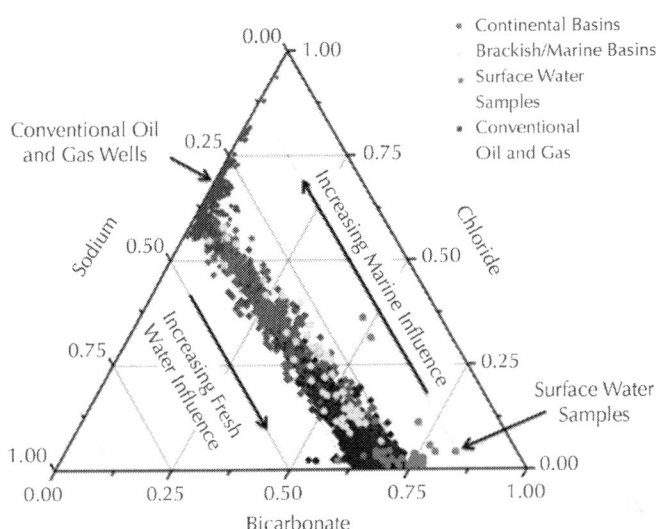

Figure 6: Ternary plot of sodium, chloride, and bicarbonate including additional samples from CBM produced water, surface water from the Powder River Basin, and conventional oil and gas wells. Note; Grayscale points are CBM produced water from the database as previously shown in Fig. 5.

The Na–Ca–Mg ternary diagram consistently exhibits a gradient linked to recharge proximity; however, this trend is limited in extent for conventional well water quality as most wells exhibit water compositions dominated by sodium (Fig. 5). The full extent of aquifer connectivity is unknown for the CBM basins due to limited information on the subject, which was highlighted in a recent National Research Council (NRC) report as a key area for research (National Research Council 2010). The surface water samples from the Powder River Basin are the only samples within the diagram that exhibit calcium and magnesium dominance (Fig. 5). The relative abundance of sodium, calcium and magnesium in well samples may act as indicators of recharge pathways and indicate appropriate wells for treatment and beneficial use.

Conventional oil and gas samples are recognized as Na–Cl-dominated water with very little bicarbonate influence and represent marine origins (Fig. 6). Bicarbonate-dominated TDS in CBM produced water is linked to both the coalification process and to sulfate reduction (Taulis, 2007). This pathway is supported by this ternary diagram where conventional oil and gas basins generally do not overlap the CBM wells. Continental CBM basins with fresh water influences generally plot on the opposite end of the spectrum inFig. 6, while brackish or marine CBM basins span the range of water compositions between Na–HCO$_3$ and Na–Cl. Surface water samples do not consistently follow the same Na–Cl–HCO$_3$ distribution as those waters are not necessarily dominated by these three ions. Trends of increasing marine and freshwater influences on the fractions for sodium, chloride, and bicarbonate are clearly shown on the composite diagram including surface waters, CBM produced water from multiple basins, and conventional oil and gas wells.

Trends in Minor Ions

Based on an evaluation of the database, constituents present at elevated concentrations due to brackish-water depositional environments include chloride, bromide, boron, TDS, SAR and trace metals. Produced waters from marine basins are expected to have higher boron, chloride, and trace metal concentrations than produced waters from continental depositional environments (Bouška, 1981 and Cheung et al., 2009). Trace metals represent a higher fraction of the minor ions in the Raton Basin (62%) than the Powder River Basin (6%) due to the marine influence in the depositional environment (Supplementary Data Fig. 1). The average trace metal fraction in the San Juan Basin is 33% due to contributions from both fresh and saline depositional environments.

Molybdenum, lead and manganese were among the trace elements exceeding beneficial use requirements in a majority of CBM produced water samples in the database (Dahm et al., 2011). Aquifers with coal beds formed under brackish and marine conditions are expected to have higher concentrations of these constituents than aquifers within coal beds formed from continental deposits (Cheung et al., 2009). Although trace metals and boron are not commonly measured in CBM databases, chloride is a major ion commonly reported with high levels associated with brackish or marine depositional environments. Additional research is needed to trace ion specific dissolution along geochemical flow paths in order to develop specific chemical reference trends by individual trace metal.

Beneficial use of Produced Water

The oil and gas sector has the opportunity to participate in the management of water for beneficial use in producing regions. As a water producer the industry has the potential to supply water for stream flow and reservoir augmentation, aquifer storage and recharge, livestock, irrigation, industrial, and municipal purposes. Beneficial use of produced water from CBM production is influenced by water quality, which based on study observations includes aquifer recharge, groundwater mineral interactions, and depositional environment. Generally the following attributes are characteristic of each source of variability:

Aquifer Recharge
- Close proximity to fresh water results in decreased TDS and trace metals concentrations
- Active recharge dilutes depositional environment characteristics

Groundwater Mineral Interactions
- Cation exchange along the flow path increases the SAR
- Natural mineral dissolution increases dissolved solids concentrations

Depositional Environment
- Continental depositional environment reflect surface water deposits with lower TDS
- Brackish/marine depositional environments exhibit higher TDS, chloride, and trace metal concentrations

The composite geochemical database analyzed provided comprehensive information on water quality characteristics and trends for over 3000 wells in the predominant CBM production areas in the western US. This information can be utilized in designing treatment processes, informing discharge permit regulations, and matching source water qualities to desired chemical characteristics for beneficial use. Understanding water quality variability is vital to designing technology that is specific to the CBM produced water composition at a potential treatment site. Treatment design based on the entire range of produced water quality is more expensive than designing treatment for specific locations based on identifiable water composition.

Water quality variability originating from aquifer water dilution from groundwater recharge, subsurface ion exchange and microbial processes, and from the coal depositional environment indicate likely environments for beneficial use opportunities. Beneficial use opportunities coincide with these attributes as close proximity to recharge and low residence time relates to fresher water more appropriate for drinking. Meanwhile less stringent requirements for livestock water make brackish, lower temperature water appropriate for use. The use of commonly measured constituents allows application of these relationships to other existing databases to infer information about water composition, thus increasing the potential to design treatment facilities specific to the chemical conditions at the specific well site. Relationships identified in this study highlight water quality compositions matching opportunities for beneficial use.

SUMMARY AND CONCLUSIONS

The opportunity to utilize CBM produced water as a resource depends on the ability to determine suitable water quality for treatment design and beneficial use. This investigation revealed that the most prevalent constituents in this water are a result of the depositional environment and mixing with fresh water recharge, confirming the hypothesis that water quality is an indication of groundwater connection. Furthermore, these analyses show that key characteristics in describing connections based on water quality in an aquifer are 1) the presence of the dominant cation sodium originating through depositional history and mineral exchange, 2) the relative absence of the major cations calcium and magnesium through mineral precipitation, and 3) the presence of the major anion chloride due to residual concentrations from marine deposits and mineral dissolution.

Specific conclusions from this study are:

- PCA analysis of the database identified three components that correlated to attributes of the study basins: Aquifer recharge (PC-1) representing 36.9%, groundwater–mineral interactions (PC-2) representing 13.8% and depositional environment (PC-3) representing 11.2% of the variations in CBM produced water quality.

- Ternary diagrams of Na–Ca–Mg and Na–HCO$_3$–Cl indicate relationships between well proximity to recharge and depositional environment formation chemistry, respectively.

- Water chemistry indicating a fresh water recharge signature was not observed in CBM wells from this study, suggesting that wells were not close enough in proximity to source areas to exhibit freshwater recharge characteristics of high calcium and magnesium to sodium ratios (Fig. 3).

- Brackish or marine depositional environments in the basins studied can be identified by the unique ternary relationship created between sodium, chloride and bicarbonate, with increasing chloride fractions indicating marine influence (Fig. 4).

- General trends for water quality characteristics of coal depositional environments were confirmed with 139 additional well samples from CBM basins worldwide. Trends were confirmed to be uniquely characteristic to CBM producing basins through comparison to over 758 conventional oil and gas wells and surface water samples from the Powder River Basin (Fig. 5 and Fig. 6).

- Improved understanding of produced water quality compositions that are appropriate for beneficial use aids the energy industry in exploring the new frontier of water resource production.

ACKNOWLEDGMENTS

This research was funded by the Research Partnership to Secure Energy for America (RPSEA) program administered by the National Energy Technology Laboratory of the U.S. Department of Energy (DOE). The authors thank the agency for funding this project to determine the beneficial use potential and best management strategies for coalbed methane produced water resources in the Rocky Mountain region. Additionally, the authors thank the regional participating CBM producers for their technical assistance, support, and contributions of information for this study.

REFERENCES

1.   Bethke, C.M., Johnson, T.M., 2002. Paradox of groundwater age. Geology 30, 385e 388.

2.   Bouska, V., 1981. Geochemistry of Coal. Elsevier Scientific Pub. Co.

3.   Bureau of Reclamation, 2003. Water 2025: preventing crisis and conflict in the West. In: Department of the Interior.

4.   Cheung, K., Sanei, H., Klassen, P., Mayer, B., Goodarzi, F., 2009. Produced fluids and shallow groundwater in coalbed methane (CBM) producing regions of Alberta, Canada: trace element and rare earth element geochemistry. Int. J. Coal Geol. 77, 338e349.

5.   Clark, C.E., Veil, J.A., 2009. Produced Water Volumes and Management Practices in the United States. ANL/EVS/R-09/1.

6.   Cooper, J.R., Crelling, J.C., Rimmer, S.M., Whittington, A.G., 2007. Coal metamorphism by igneous intrusion in the Raton

Basin CO and NM: Implications for generation of volatiles. Int. J. Coal Geol. 71, 15e27.

7. Dahm, K.G., Guerra, K.L., Xu, P., Drewes, J. r. E., 2011. Composite geochemical database for coalbed methane produced water quality in the rocky mountain region. Environ. Sci. Technol. 45, 7655e7663.

8. Freeze, R.A., Cherry, J.A., 1979. Groundwater. Prentice-Hall International, Hemel Hempstead.

9. Garrels, R.M., Christ, C.L., 1965. Solutions, Minerals, and Equilibria. Harper & Row, New York.

10. Jackson, R.E., Reddy, K.J., 2007. Geochemistry of coalbed natural gas (CBNG) produced water in the powder river basin, Wyoming: salinity and sodicity. Water Air Soil Pollut. 184, 49e61.

11. Johnston, C.R., Vance, G.F., Ganjegunte, G.K., 2008. Irrigation with coalbed natural gas co-produced water. Agric. Water Manag. 95, 1243e1252.

12. Kinnon, E.C.P., Golding, S.D., Boreham, C.J., Baublys, K.A., Esterle, J.S., 2010. Stable isotope and water quality analysis of coal bed methane production waters and gases from the Bowen Basin, Australia. Int. J. Coal Geol. 82, 219e231.

13. Konhauser, K., 2007. Introduction to Geomicrobiology. Blackwell Science.

14. Larson, L.R., Daddow, R.L., 1984. Ground-water-quality data from the Powder River Structural Basin and Adjacent Areas, Northeastern Wyoming. U.S. Geological Survey.

15. Orem, W.H., Tatu, C.A., Lerch, H.E., Rice, C.A., Bartos, T.T., Bates, A.L., Tewalt, S., Corum, M.D., 2007. Organic compounds in produced waters from coalbed natural gas wells in the Powder River Basin, Wyoming, USA. Appl. Geochem. 22, 2240e2256.

16. Pashin, J.C., 2007. Hydrodynamics of coalbed methane reservoirs in the Black Warrior Basin: key to understanding reservoir performance and environmental issues. Appl. Geochem. 22, 2257e2272.

17. Rice, C.A., 2003. Production waters associated with the Ferron coalbed methane fields, central Utah: chemical and isotopic composition and volumes. Int. J. Coal Geol. 56, 141e169.

18. Rice, C.A., Flores, R.M., Stricker, G.D., Ellis, M.S., 2008. Chemical and stable isotopic evidence for water/rock interaction and biogenic origin of coalbed methane, Fort Union Formation, Powder River Basin, Wyoming and Montana U.S.A. Int. J. Coal Geol. 76, 76e85.

19. K.G. Dahm et al. / Journal of Cleaner Production 84 (2014) 840e848 847 Rice, C.A., Nuccio, V., 2000. Water Produced with Coal-Bed Methane. In: United States Geological Survey. Department of the Interior, Denver, Colorado.

20. Rice, D.D., Clayton, J.L., Pawlewicz, M.J., 1989. Characterization of coal-derived hydrocarbons and source-rock potential of coal beds, San Juan Basin, New Mexico and Colorado, U.S.A. Int. J. Coal Geol. 13, 597e626.

21. Taulis, M., 2007. Groundwater characterization and Disposal Modelling for Coal Seam Gas Recovery. University of Canterbury, Christchurch, New Zealand.

22. United States Geological Survey, 2002. In: Produced Waters Database. U.S. Department of the Interior.

23. Van Voast, W.A., 2003. Geochemical signature of formation waters associated with coalbed methane. AAPG Bull. 87, 667e676.

8

Characterization of a Methanogenic Consortium Enriched From a Coalbed Methane Well in the Powder River Basin, U.S.A.

Michael S. Green, Keith C. Flanegan, and Patrick C. Gilcrease

Department of Chemical and Biological Engineering, South Dakota School of Mines and Technology, Rapid City, SD, United States

ABSTRACT

Well-bore water samples from the Fort Union Formation in the Powder River Basin of Wyoming tested positive for the presence of living microbial communities capable of generating methane from Wyodak coal under laboratory conditions. The methanogens in this consortium produced methane from acetate and methanol, but did not produce methane from a H_2-CO_2 headspace. This was consistent with a phylogenetic analysis of archaeal 16S ribosomal deoxyribonucleic acid (rDNA) sequences from the enrichment culture, which revealed just two phylotypes, both closely related to *Methanosarcina mazei*.Phylogenetic analysis of bacterial 16S ribosomal ribonucleic acid (rRNA) genes revealed phylotypes similar to the acetogens *Acidaminobacter hydrogenoformans* and *Syntrophomonas* sp., and to known fermentative species. This methane-producing consortium was maintained on a defined microbial medium supplemented with Wyodak coal plus 50 mg/L yeast extract as the sole carbon substrates. At 22 °C, the maximum methane production rate was 0.084 m³/t coal/day (2.7 scf/ton/day); in comparison, total methane reserves in the Powder River Basin are approximately 1.6 to 2.2 m³/t (50 to 70 scf/ton). When the incubation temperature was increased from 22 °C to 38 °C, the rate of methane production increased by 300%; similarly, when the culture medium pH was lowered from 7.4 to 6.4, the methanogenesis rate increased by 680%. Increasing the coal particle surface area by 890% via smaller particle size increased methane production rates by over 200%. Microbial methane production in coal slurries was also enhanced by the addition of the solvent *N,N*-dimethylformamide (DMF). These results suggest an opportunity to enhance coalbed methane reserves by stimulating the activity of existing methanogenic consortia *in-situ*; in particular, reservoir treatments that enhance coal solubility and dissolution rates may be beneficial.

INTRODUCTION

History of Coalbed Methane in the Powder River Basin

Interest in the development of methane resources in the Powder River Basin (PRB) of Wyoming (see Fig. 1for map of basin) began in 1957 when the U.S. Geological Survey (USGS) reported that methane had been detected from water wells in the area (Flores, 1999 and Flores, 2004). Commercial exploration and production began in the early 1980s (Flores, 1999); gas-in-place in the PRB has been estimated at 1.5 trillion m^3 (51 trillion cubic feet) (De Bruin et al., 2000). While the concentration of gas in PRB seams (1.6 to 2.2 m^3/t coal) is low when compared to other U.S. coalbed methane reserves (up to 22 m^3/t), the shallow depth of the coal seam (< 300 m) and favorable natural gas prices have made production economical (Mavor et al., 1999 and De Bruin et al., 2000). Annual production from the PRB has increased 200% since 2001 to 9.5 billion m^3 in 2005, which represents 20% of the U.S. coalbed methane production. While most natural gas deposits found in limestone or sandstone contain only 70–80% methane, coalbed gas in the PRB is typically 97% methane and H$_2$S free, making production simple and cost-effective (Fellows Energy, 2005). A typical completed well costs around $165,000, and will begin to produce gas after a dewatering period of up to six months (Swindell, 2007). Dewatering may be detrimental to indigenous microbial populations, but the real consequences of this production practice on future *in-situ* biogenic gas production are not known at this time.

Methane is generated during coalification in both biogenic and thermogenic processes. While primary biogenic methane is generally lost, meteoric recharge can transport microbes into mature coal seams, resulting in the formation of secondary biogenic methane (Scott et al., 1994 and Faiz and Hendry, 2006). Isotope analyses of coalbed methane from the Powder River Basin indicate that it is almost completely of biogenic origin (Gorody, 1999, Faiz and Hendry, 2006 and Flores et al., 2008-this volume). The low rank, high permeability, and high water content of PRB coal suggests that seam conditions are still favorable for methane biogenesis (Rice, 1993). If *in-situ* methane biogenesis reactions are indeed active, a fundamental understanding of these processes could lead to enhanced exploration methods, as well as enhanced methane production.

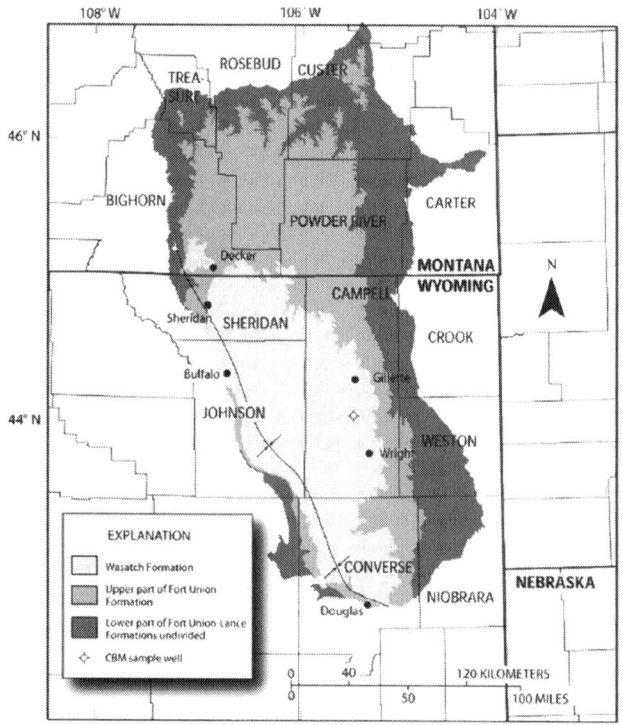

Figure 1: Map of the Powder River Basin, showing location of the sampled well.

Microbiology of Methanogenesis

Methanogens are strictly anaerobic microorganisms which produce methane gas as their end product of metabolism. They can only use a limited number of simple carbon compounds as substrates; the most common substrates are H_2–CO_2 and acetate. Methanogens that utilize methanol, methylamines, and dimethyl sulfide are known as methylotrophs (Zinder, 1993). Formate along with a few other alcohols (ethanol and isopropanol) can also be oxidized by some methanogens (Widdel, 1986). For the conversion of complex organic substrates (such as coal) to methane, fermentative and acetogenic bacteria are required in addition to methanogens. First, fermentative bacteria hydrolyze and then ferment complex substrates to produce acetate, longer chain fatty acids, carbon dioxide, H_2, NH_4^+, and HS^-. H_2-using acetogenic bacteria consume H_2 and carbon dioxide to produce more acetate; in addition, they can demethoxylate low-molecular-weight ligneous materials and ferment some hydroxylated aromatic compounds to produce acetate (McInerney and Bryant, 1981). H_2-producing acetogens convert fatty acids, alcohols, and some aromatic and amino acids to the H_2, carbon dioxide, and acetate needed by the methanogens. This collection of different microbial species is referred to as a consortium; for methanogenic consortia, interdependencies such as interspecies H_2 transfer are common (Zinder, 1993).

Concept of Microbially Enhanced Coalbed Methane

Scott (1999) defines the concept of Microbially Enhanced Coalbed Methane (MECoM) as the introduction of bacterial consortia and nutrients into coal beds; this not only has the potential to produce new methane from the coal, but could also increase reservoir permeability via the microbial consumption of coal, waxes, and paraffins (Scott, 1999). The introduction of a bacterial consortium may not be necessary, as indigenous microbial populations may already be present (Faison, 1992). Biodegradation studies frequently show that contaminated sites will select for organisms that can degrade the organic compounds present, and these indigenous populations will perform as well or better than any foreign microbes that are introduced to the site (Baker, 1994 and Cookson, 1995). In-situ bioremediation strategies typically involve the stimulation of indigenous organisms via the introduction of a limiting nutrient(s) (Skladany and Baker, 1994). A number of laboratory studies have demonstrated that bacterial consortia are capable of converting coal into methane (Harding et al., 1993, Volkwein et al., 1994 and LUCA Technologies, LLC, 2004). This is not surprising, as low-rank coals are enriched in low-molecular-weight, leachable organic compounds that may be amenable to microbial degradation (Faison, 1992).

Concept of Mass Transfer Limited Bioconversion

While coal does contain biodegradable compounds, its persistence in the environment over millions of years indicates that there are significant resistances to coal biotransformation in nature (Faison, 1992). Low-rank coals have a lignin-like polymer structure that is recalcitrant to biodegradation, particularly under anaerobic conditions (Faison, 1992 and Fakoussa and Hofrichter, 1999). In addition, some of the organic compounds and metals present in coal may be inhibitory to microbial growth (Faison, 1992). As microorganisms are inherently aquatic, they need to be associated with an aqueous phase to survive (Prescott et al., 1990); thus, substrates need to be dissolved in aqueous solution for microbial uptake (Ogram et al., 1985, Bailey and Ollis, 1986 and Ramaswami and Luthy, 1997). This mass transfer process involves dissolution of the substrates at the coal–water interface followed by diffusion through the aqueous phase to the degrading microbes. Studies have shown that overall biodegradation rates for organic solids can be limited by the solid–liquid mass transfer rate when the microbial concentration is high and/or the solid surface area is low (Gilcrease, 1997); as such, the rate and quantity of methane generated from coal may depend on the exposed surface area, and the insolubility and impermeability of coal represent major constraints. Although the diameter of coal pores ranges from 0.04 to 30 µm, many of these pores are too small to permit microbial entry (< 0.2 µm); thus, microbial access is mostly limited to the cleat surfaces of the coal (Faison, 1992 and Scott, 1999).

Goals of This Study

The goals of this study were as follows: (1) demonstrate that viable methanogenic consortia are present in PRB coalbed methane seams, and are capable of using coal as the primary substrate; (2) determine which intermediates are utilized by the methanogens present in the consortium; (3) use a phylogenetic analysis of bacterial and archaeal 16S rRNA genes to characterize different members of the microbial consortium; (4) evaluate the effects of temperature, pH, and particle size on biomethane production from coal; and, (5) determine if the system is mass transfer limited, and if so, evaluate solvents for enhancing coal mass transfer rates.

MATERIALS AND METHODS

Detection of Methanogens

An enrichment culture technique was used for detecting and characterizing a methanogenic consortium present in a well-bore water sample. Water samples were transferred to a nutrient medium containing coal and/or standard methanogenic substrates such as acetate or a H_2–CO_2 headspace. When viable methanogens are present in the sample, culture growth and concomitant methane production will ensue; thus, a headspace analysis for methane production serves as a positive identification for methanogens in a sample when compared to an autoclaved control. This technique has two limitations: (1) the methanogens in the sample must remain viable during the sampling and culturing process, and (2) the proper substrates for the particular methanogens present must be provided in the medium. Because of the first limitation, particular care must be taken to ensure strict anaerobic conditions throughout the sampling and culturing process.

Inoculum Source

A well water sample was collected in May 2002 from the non-producing Carter Federal #8L well owned by Peabody Natural Gas (API: 49-005-34969). This well is located in Campbell County, Wyoming, Section 8, Township 47 N, and Range 72 W (see Fig. 1), and is completed into the Fort Union Formation (see Flores et al., 2008-this volume for stratigraphic position). The total measured well depth was 237 m (778 ft), and depth to the water level was 225 m (737 ft). Estimated pressure at the bottom of the well was 1410 kPa (190 psig) (Mavor et al., 1999). The collection device consisted of a 1-m long, 10-cm diameter polyvinyl chloride pipe that was sealed at the bottom and capped with an open steel mesh at the top. Both pipe and mesh enclosure were sterilized with a bleach solution at the field site and then lowered to the bottom of the well on a wire line. Upon immersion, well-bore water entered the collection device through the top mesh enclosure. Once the filled collection device had been retrieved from the well-bore, the mesh cap was removed and the sample water was aseptically transferred to sterile glass jars that had been sparged with nitrogen. Due to the physical separation of the well-bore from the native coal seam, the collected consortium may or may not be representative of the microbes that are active in the coal seam. The water temperature was 17.5 °C, as measured in the field immediately after sampling. Field analysis of the water sample using test strips (WaterWorks™) showed a pH of 8.0, a total hardness (as calcium carbonate) of 180 ppm, a total alkalinity of $\geq$ 360 ppm, NaCl $\leq$ 1500 ppm, and no detectable (less than 1 ppm) iron, nitrate, or nitrite.

Coal Substrate

For cultures containing coal, subbituminous B Wyodak was used as the primary carbon source. This coal was obtained in 2004 from the Department of Energy Coal Sample Bank at Pennsylania State University (Sample code DECS-26), where it had been stored and shipped under argon atmosphere. The moisture content of the coal was 26%, and the elemental analysis on a dry-basis free was 69.74% C, 5.55% H, 0.94% N, 0.35% organic S, 14.83% O, and 8.59% mineral matter (including 0.13% FeS_2). Upon receipt by our laboratory, the coal was exposed to air for several hours during sieve classification; each size fraction was then collected and sealed under nitrogen atmosphere until used in culture experiments. Unless noted otherwise, a 30–60 mesh size fraction of coal (250–600-µm particle diameter) was used in the culture experiments.

Medium

The growth medium used in these experiments was adapted from a recipe by Tanner (2002) for culturing strict anaerobes. The medium was prepared in 100-mL batches with distilled water. The final concentration of minerals in the medium (g/L) were as follows: NaCl, 0.8; NH_4Cl, 1.0; KCl, 0.1; KH_2PO_4, 0.1; $MgSO_4·7H_2O$, 0.2; $CaCl_2·2H_2O$, 0.04. The final concentration of trace metals in the medium (mg/L) were: nitrilotriacetic acid, 10; $MnSO_4·H_2O$, 5; Fe $(NH_4)_2(SO_4)_2·6H_2O$, 4; $CoCl_2·6H_2O$, 1; $ZnSO_4·7H_2O$, 1; $CuCl_2·2H_2O$, 0.1; $NiCl_2·6H_2O$, 0.1; $Na_2MoO_4·2H_2O$, 0.1; Na_2SeO_4, 0.1; Na_2WO_4, 0.1. Vitamin concentrations (mg/L) in the final medium were: pyridoxine·HCl, 0.1; thiamine·HCl, 0.05; riboflavin, 0.05; calcium pantothenate, 0.05; thioctic acid, 0.05; p-aminobenzoic acid, 0.05; nicotinic acid, 0.05; vitamin B_{12}, 0.05; mercaptoethanesulfonic acid (coenzyme M), 0.05; biotin, 0.02; folic acid, 0.02. Yeast extract (50 mg/L) was added to provide undefined growth factors, and N-[Tris(hydroxymethyl)methyl]-2-aminoethanesulfonic acid (TES) (2.0 g/L) was added as a pH buffer. After adjusting the pH of the medium to 7.2 with sodium hydroxide, sodium bicarbonate (2.0 g/L) was added. A drop of resazurin was added as an oxygen indicator (resazurin has a pink color at redox potentials above about − 150 mV), and the 100 mL solution was then boiled for 60 s under a nitrogen purge to remove dissolved oxygen. After cooling under a nitrogen flow, the solution was transferred to an anaerobic chamber (Coy 7150-000) where 0.5 mL of reducing solution was added; the reducing solution contained (g/L): NaOH, 9; $Na_2S·9 H_2O$, 40; l-cysteine·HCl hydrate, 40. The medium was then dispensed inside the chamber in 10-mL aliquots to Balch tubes (Bellco 2048-00150) containing 0.25 g of Wyodak coal, sealed with butyl rubber stoppers (Bellco 2048-11800), and then removed from the chamber. Outside the chamber, a gassing manifold with needle tips (Balch and Wolfe, 1976) was used to vacuum and fill the headspace with 300 kPa nitrogen for three cycles. Prior to inoculation, the sealed, pressurized tubes were sterilized in an autoclave at 121 °C for 15 min. Final medium pH after autoclaving was 7.3.

For methanogenic substrate screening studies, sodium acetate trihydrate (5.0 g/L) or methanol (4.0 g/L) or a 300 kPa headspace of 80% H_2–20% CO_2 were provided in lieu of a coal substrate. For the solvent addition experiments (Section 3.8), solvents were added to the culture tubes after the gassing manifold step but prior to autoclaving.

Inoculum Maintenance

Raw well water stored at room temperature under anaerobic conditions was used as the inoculum in culture experiments designed to screen for methanogenic substrates, nutrient amendments, and lignin-derived intermediates (3.1. Substrate and nutrient results section). A methanogenic consortium enriched from this well water sample was maintained in the laboratory at 30 °C on the Tanner medium described above, with 25 g/L coal and 50 mg/L yeast extract as the sole carbon-energy sources. One milliliter of this consortium was transferred to 10-mL fresh medium every six weeks (approximately). All other experiments were inoculated from this maintenance culture line. It is likely that this maintenance technique enriched for selected species in the original consortium; as such, the results are not necessarily representative of the metabolism of the native coal bed consortium. Also, changes in the microbial population in the maintenance culture over time may explain why cultures run at the same set of conditions at different times yielded different methane production curves. For culture transfers, a flame-sterilized Hungate probe was used to fill sterile glass syringes with nitrogen prior to piercing the septa of experimental tubes, assuring that no oxygen was introduced. The aseptic, strict anaerobic techniques used were originally described by Hungate (1969) and modified by Balch and Wolfe (1976).

Growth Conditions

At time zero, 10 mL of culture medium was inoculated with 1 mL of raw well water or 1 mL of the maintenance culture. All cultures were grown in sealed Balch tubes under a 300-kPa nitrogen atmosphere. These tubes were placed horizontally in an incubator shaker at constant temperature (30 °C unless otherwise noted) and agitated at 100 rpm to maximize coal-liquid mass transfer rates. Cultures were removed from the shaker only for short sampling periods. Cell concentrations were not quantified, as the presence of coal solids would have interfered with such measurements. For killed cell controls, the inoculum source was autoclaved for 15 min at 121 °C prior to transfer to the culture tubes.

Microscopy

Aqueous culture samples were examined using an Olympus CX41 phase-contrast light microscope at 400× and 1000×. Coal solids were harvested from cultures, dried at 37 °C overnight, and examined with a Zeiss Supra40 VP field-emission SEM at 17,000× to 120,000×; samples were imaged without conductive coatings with the secondary electron detector at an acceleration voltage of 1 keV.

Gas Analysis

Culture tube headspace samples were drawn using aseptic, anaerobic technique with a 100 µL gas syringe (Hamilton 81056) equipped with a shut-off valve (Hamilton 35083) and a sterile 23 gauge needle (BD 305193); the shut-off valve allowed for analysis of the sample at the original headspace pressure. Methane analysis was performed using a Hewlett-Packard 6890 gas chromatograph, equipped with a Chrompack Carboplot P7 25-m × 0.53-mm fused silica capillary column (HP 19095P-CO2) and a thermal conductivity detector. The helium (carrier gas) flowrate was set at 5 mL/min, with a 4.5:1 split ratio. The injection port was maintained at 200 °C, the oven temperature was 105 °C, and the TCD operated at 250 °C. The retention time for methane was 2.9 min. Calibration standards consisting of 4% or 10% methane (Matheson Tri-Gas GMT10404TC, GMT10320TC, respectively) were injected at atmospheric pressure to generate the calibration plot. Methane–water and methane–coal equilibrium calculations were used to demonstrate that headspace methane levels are representative of the total amount of methane produced by the cultures. A Henry›s law constant for methane in water at 25 °C (4185 MPa) (Smith et al., 2005) was used to estimate the maximum amount of methane dissolved in the aqueous culture medium; for a maximum methane headspace level of 0.05 mmol per 16 mL of total headspace at 300 kPa (see Fig. 5), the dissolved methane level was estimated at < 0.01% of the headspace methane level. Similarly, for a methane Langmuir pressure of 2717 kPa and a methane Langmuir storage capacity of 2.4 m³/t at 18 °C (Mavor et al., 1999), total methane adsorbed to the coal present in our cultures was estimated at < 0.2% of the headspace methane level.

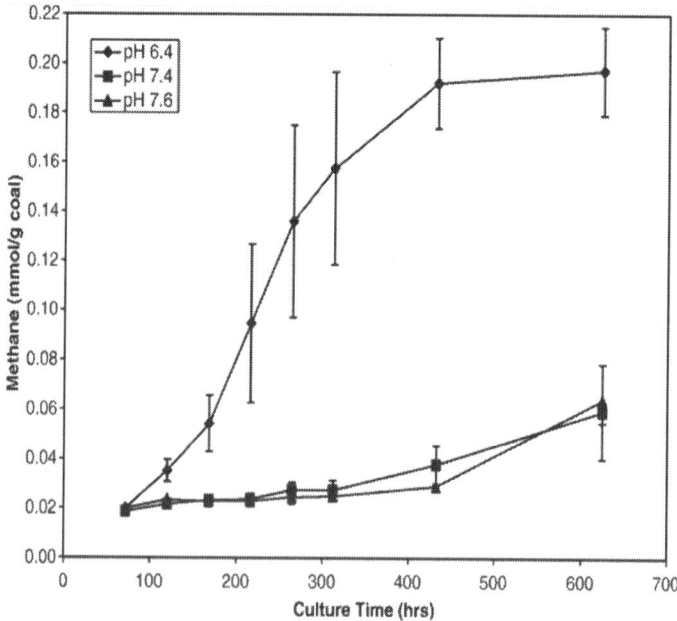

Figure 2: Microbial methane production from Wyodak coal at 30 °C for pH 6.4, 7.4, and 7.6. Each culture contained 0.25 g of 30–60 mesh Wyodak coal. Error bars represent 1 standard deviation for triplicate cultures. No methane was observed in killed cell controls.

DNA Extraction, Library Construction, and Phylogenetic Analyses

Detailed information regarding the composition of a microbial community can be obtained from a phylogenetic analysis of 16S rDNA sequences; this technique avoids certain biases associated with cultivation studies (Amann et al., 1995). The microbial consortia in anaerobic digesters (Godon et al., 1997 and Leclerc et al., 2004), lake sediments (Wani et al., 2006), and an acidic peat bog (Bräuer et al., 2006) have been analyzed using this approach. DNA was extracted from 10 mL of a 22-day-old maintenance culture using a Power soil DNA kit (MO Bio, Carlsbad CA) according to the manufacturer's instructions. Retrieved DNA was used as the template in PCR amplification. Bacterial 16S rRNA genes were amplified using a *Bacteria*-specific primer set: 530F (5'-GTCCCAGCMGCCGCGG-3') and 1490R (5'-GGTTACCTTGTTACGACTT-3') (Wani et al., 2006); this primer set has been widely used by several researchers for preparation of 16S rDNA libraries (Sekiguchi et al., 1998, Rossetti et al., 2003 and Wani et al., 2006). A second *Archaea*-specific primer set was also used: S-D-Arch-0021-a-A-20 (5'-TTCCGGTTGATCCYGCCGGA-3') and S-D-Arch-0958-a-A-19 (5'-YCCGGCGTTGAMTCCAATT-3') (Wani et al., 2006). It has been shown that partial sequences containing variable regions produce much the same phylogenetic tree structures as full length 16S rRNA gene sequences (Borneman et al., 1996 and Kemp and Aller, 2004). Amplification was carried out in a 50-µL reaction mixture containing 1.5 mM $MgCl_2$, 0.2 mM each dNTPs, 25 pmol of forward and reverse primer, 50 ng DNA template, and 1 U*Taq* DNA polymerase (New England Biolab, MA) with reaction buffer supplied by the manufacturer. Hot-start PCR was performed at 95 °C prior to addition of DNA *Taq* polymerase; 30 cycles of 45 s at 94 °C, 60 s at 50 °C and 90 s at 72 °C, followed by a final extension of 10 min at 72 °C. A negative control (PCR reaction without a DNA template) was included in all PCR experiments to check for any non-specific amplification. PCR products were cloned in pGEM-T Easy vector and transformed into

*Escherichia coli*JM109 High-Efficiency Competent Cells (Promega, Madison, WI) as per manufacturer's protocol. Plasmids were isolated using a plasmid extraction kit (Qiagen, CA) and nucleotide sequences of cloned genes were determined by sequencing with a M13F (5'-GTAAAACGACGGCCAG- 3') primer in an automated 3730 DNA analyzer (Applied Biosystems, CA). Sequencing was carried out for 37 and 49 randomly selected correct-sized clones from bacterial and archaeal 16S rDNA libraries, respectively. The presence of chimeric sequences was checked using the check-chimera program (Maidak et al., 1997). A similarity search for sequences was carried out by BLAST (Altschul et al., 1990) and alignment was carried out by CLUSTAL W (Thompson et al., 1994). Ambiguously aligned portions were deleted by DAMBE (Xia, 2000). A Jukes–Cantor corrected distance matrix was constructed using the DNADIST program of PHYLIP (Felsenstein, 1989). This distance matrix was used as an input file to assign sequences in various operational taxonomic units (OTUs) using the furthest neighbor algorithm of DOTUR (Schloss and Handelsman, 2005). In our study, sequences which were 97% (0.03 evolutionary distance) identical were assigned to the same OTU (Wani et al., 2006). Bootstrap resampling analysis for 1000 replicates was performed to estimate the confidence of tree topology. The phylogenetic tree was constructed by the Neighbor-joining method (Saitou and Nei, 1987) using the Jukes–Cantor nucleotide substitution model by MEGA v3.1 (Kumar et al., 1993).

Accession Numbers

All sequences from this study were deposited in GenBank. Bacterial sequences were submitted under accession numbers EU071185, EU071186, EU071187, EU071188, EU071189, EU071190, EU071191, EU071192, EU071193, EU071194, EU071195, EU071196, EU071197, EU071198, EU071199, EU071200, EU071201, EU071202, EU071203, EU071204, EU071205, EU071206, EU071207, EU071208, EU071209, EU071210, EU071211, EU071212, EU071213, EU071214, EU071215,

EU071216, EU071217, EU071218, EU071219, EU071220 and EU071221. Archaeal sequences were assigned accession numbers EU071136, EU071137, EU071138, EU071139, EU071140, EU071141, EU071142, EU071143, EU071144, EU071145, EU071146, EU071147, EU071148, EU071149, EU071150, EU071151, EU071152, EU071153, EU071154, EU071155, EU071156, EU071157, EU071158, EU071159, EU071160, EU071161, EU071162, EU071163, EU071164, EU071165, EU071166, EU071167, EU071168, EU071169, EU071170, EU071171, EU071172, EU071173, EU071174, EU071175, EU071176, EU071177, EU071178, EU071179, EU071180, EU071181, EU071182, EU071183 and EU071184.

RESULTS

Substrate and Nutrient Results

Cultures inoculated with sampled coalbed methane well-bore water tested positive for methane production, while no methane was observed for killed cell controls. This methanogenic consortium has consistently demonstrated methane production when Wyodak coal (plus 50-mg/L yeast extract) is the sole carbon and energy source provided. A screening of known methanogenic substrates showed that methanogens present in this consortium can utilize acetate (37 mM) or methanol (124 mM or 0.5 vol.%) to produce methane, while no methane production was observed when a H_2–CO_2 atmosphere (300 kPa, 80 mol% H_2, 20 mol% CO_2) was provided as the carbon-energy source.

To evaluate real-time methanogenesis under reservoir-like conditions, 10 mL of well water was used as the medium and 0.25-g Wyodak coal was added as the carbon-energy source. Following autoclave sterilization of the coal-water slurry, 1 mL of raw well water was added as the inoculum. After sixteen days, no methane was produced from coal when unamended well water or well water plus 50 mg/L yeast extract were used as the medium. Methane was produced when the well water was amended with other medium components (mineral solution, vitamin solution, trace metal solution, acetate) with acetate having the largest effect of the individual components. Others have identified benzoic and cinnamic acids as intermediates in the conversion of lignin and coal to methane (Colberg and Young, 1985 and Harding et al., 1993). For our cultures, the addition of cinnamic acid (20 mM) enhanced the production of methane in the presence of coal, while benzoic acid (at 20 mM) had a negative effect on methane production.

Phylogenetic Analysis of Bacterial 16s rRNA Genes

A total of 37 sequences were included in a phylogenetic analysis which generated 14 OTUs spanning a wide range within the domain *Bacteria*, occupying only three taxonomic groups. The majority of the clones were classified as belonging to *Firmicutes* (94.6%), while remaining clones were grouped into*Spirochaetes* (2.7%) and *Proteobacteria* (2.7%) (Fig. 2). From this analysis it was clear that major taxonomic lineages represented by multiple clones had been captured in our library, and there was little chance that sequencing of any additional clones would identify any new taxonomic group. While there could be some rare taxonomic lineages that were not captured in this library, it is difficult to get complete species census, and there are species that remain to be discovered in almost every molecular inventory (Curtis et al., 2006 and Schloss and Handelsman, 2006).

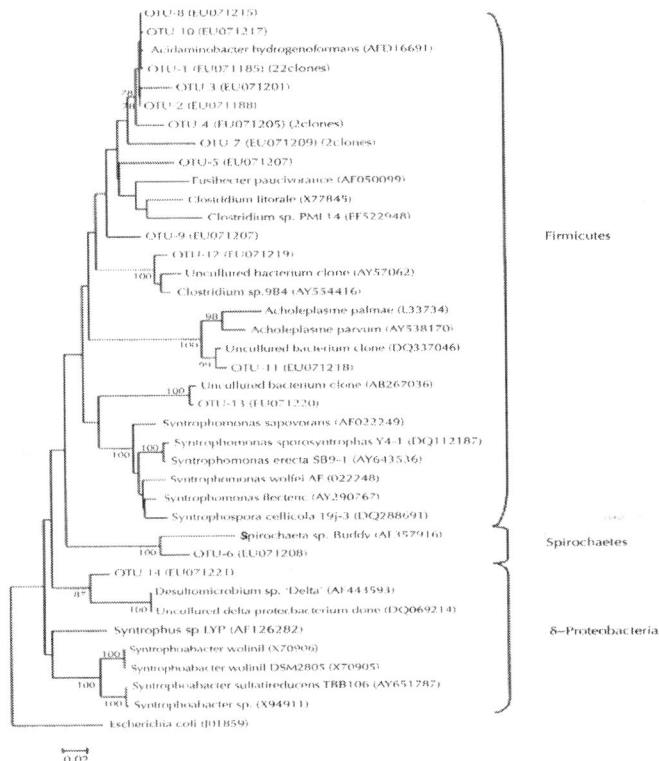

Figure 3: Phylogenetic dendrogram showing the relationship between bacterial 16S rDNA sequences with reference sequences in GenBank. For OTUs representing multiple clones, the number of additional clones is given in parentheses. *Escherichia coli* (J01859) was selected as out-group. The tree was based on Jukes–Cantor distance and constructed using a neighbor-joining algorithm with 1000 bootstrappings. The scale bar represents 0.02 substitutions per nucleotide position. Numbers at the node are the bootstrap values (%). Bootstrap values which were < 75% were not shown.

A total of 12 OTUs were clustered with cultivable members belonging to *Firmicutes*. Within the *Firmicutes*, OTUs-1, 2, 3, 4, 5, 7, 8, 9, 10, 12, and 13, containing 92% of total clone diversity were affiliated to class*Clostridia*. A cluster of 28 clones within *Clostridia* converged upon a monophyletic group with a sequence belonging to *Acidaminobacter hydrogenoformans*AF016691, an obligate anaerobe isolated from black mud (Meijer et al.,

1999). OTU-7 formed a deep branch within the *Clostridia* without affiliation to cultured genera and species or to other phylotypes. OTU-12 formed a sister lineage with AY570625, an uncultured bacterium clone reported from a low temperature biodegraded oil reservoir (Grabowski et al., 2005) and grouped together with *Clostridium* sp. 9B4 (AY554416, unpublished), an anaerobic cellulolytic bacteria from landfill leachate bioreactor supported by high bootstrap value. Genus *Clostridium* is known to produce a mixture of organic acids and alcohols from carbohydrates and peptones (Garrity, 2005). OTU-13 formed a sister lineage with an uncultured bacterium clone retrieved from upflow anaerobic sludge blanket (UASB) sludges (AB267036, unpublished) and displayed 97% DNA similarity supported by 100% bootstrap value and fell clearly in *Syntrophomonas* cluster. Members belonging to genus *Syntrophomonas* are secondary fermenters and can oxidize C_4–C_8 fatty acids, yielding the methanogenic substrates acetate, CO_2 and H_2. OTU-11 belonged to class *Mollicutes* and displayed 93% DNA similarity with an uncultured bacterium clone reported from subsurface water of the Kalahari Shield, South Africa (DQ337046, unpublished) and converged upon a monophyletic group that includes a cultured species of genus *Acholeplasma*. This genus is facultative anaerobic and can ferment carbohydrates to acids and alcohols.

Spirochaetes included the single OTU-6 (2.7% clones) that formed a common grouping with a sequence belonging to *Spirochaeta* sp. Buddy (AF357916, unpublished) and showed 87% DNA similarity with it. *Spirochaeta* related sequences have also been reported from anaerobic digesters treating wine distillation waste (Godon et al., 1997) and mesophilic and thermophilic methanogenic sludges (Sekiguchi et al., 1998). These bacteria are obligately/facultatively anaerobic and chemoorganotrophic in nature. Members of this genus ferment carbohydrates and produce acetate, succinate and lactate (Garrity, 2005). *Proteobacteria* included OTU-14 (2.7% clones), which was affiliated to -proteobacteria class and clustered around the sequence belonging to *Desulfomicrobium* sp. (AF443593, unpublished).

Phylogenetic Analysis of Archaeal 16S rRNA Genes

This phylogenetic analysis confirmed the presence of methanogens in the PRB well water enrichment culture. Of the 49 sequences analyzed, only two OTUs were generated, both affiliated to*Methanosarcinales* (Fig. 3). No OTUs related to either *Methanobacteriales* or *Methanomicrobiales* were detected. OTU-1 and OTU-2 contained 98% and 2% of total clone diversity, respectively; they were related to an uncultured archaeal clone retrieved from a low temperature biodegraded oil reservoir (AY570657, unpublished) at 97 and 98% similarity, respectively. These OTUs also converged upon a monophyletic cluster that includes the cultured acetoclastic lineage *Methanosarcina mazei* SarPi AF028691 (Joulian et al., 1998) and exhibited 96 and 97% DNA similarity with it, respectively.

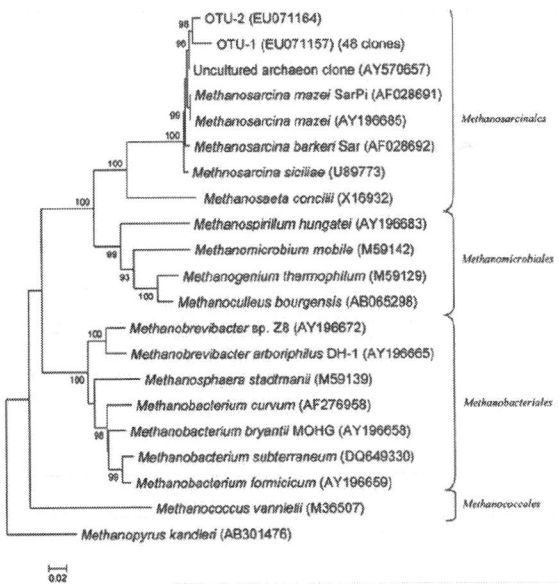

Figure 4: Phylogenetic dendrogram showing the relationship between archaeal 16S rDNA sequences with reference sequences in GenBank. For

OTUs representing multiple clones, the number of additional clones is given in parentheses. *Methanopyrus kandleri*(AB301476) was selected as outgroup. The tree was constructed as described for Fig. 2.

Microscopic Analysis

A routine maintenance culture was harvested after 19 days for microscopic analysis; at this point 0.056 mmol methane/g coal had been produced by the culture. Examination of the aqueous phase under a phase-contrast microscope at 400× and 1000× showed clusters of cocci that were somewhat asymmetrical in shape. Conversely, no microbes were apparent on the surface of coal particles harvested from the same culture when examined under a SEM at 17,000 to 120,000×.

Effect of Temperature

In this culture experiment, the biological production of methane from Wyodak coal was evaluated as a function of incubation temperature. As shown in Fig. 4, cultures grown at 30 and 38 °C had higher methane production rates from coal than was observed at 22 °C. The maximum methane production rate increased with increasing temperature, with the 38 °C tubes producing 14 µmol methane/g coal/day between 312 and 432 h while the 30 and 22 °C tubes produced 8.2 and 3.5 µmol/g coal/day, respectively, between 432 and 624 h. The final methane level in the 30 and 38 °C cultures was 118% greater than that in the 22 °C tubes, with 0.135 mmol methane/g coal being produced at each of the higher temperatures, while the 22 °C tubes produced 0.062 mmol/g coal. No detectable methane was observed for killed cell controls, which eliminates abiotic coal desorption/degradation as a potential methane source.

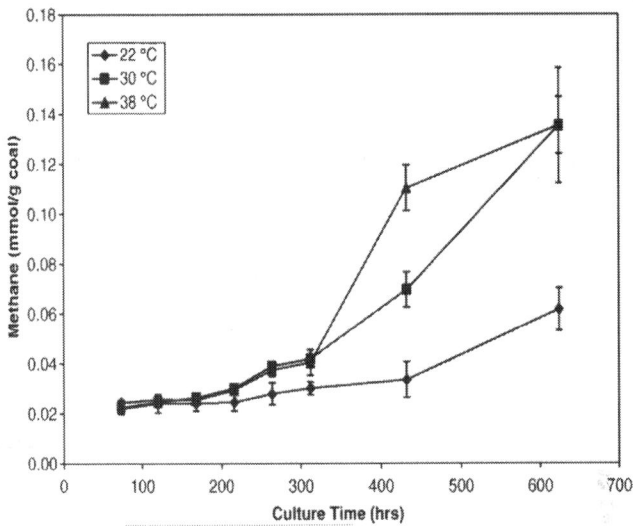

Figure 5: Microbial methane production from Wyodak coal at 22, 30, and 38 °C. Initial medium pH was 7.3 for all cultures. Each culture contained 0.25 g of 30–60 mesh Wyodak coal. Error bars represent 1 standard deviation for triplicate cultures for 22 and 38 °C, duplicate cultures for 30 °C. No methane was observed in killed cell controls.

While headspace methane levels were not measured immediately after inoculation, in a repeat experiment time zero methane concentrations were found to be negligible; thus, the methane levels at 72 h represent new methane produced and not methane transferred over with the inoculum. Methane levels in the first 72 h were independent of incubation temperature, and most likely represent the rapid metabolism of 50 mg/L yeast extract in the medium; the predicted methane yield from this yeast extract is equivalent to 0.021 mmol/g coal (assumes the yeast extract is 50% carbon (Bailey and Ollis, 1986), and that conversion of this carbon to methane is via the acetate pathway at 50% theoretical conversion (Flanegan, 2004)). It was first hypothesized that the lag phase in methane production between 72 and 168 h could have been due to competition from sulfate reducing bacteria (OTU-14 is affiliated with a known sulfate reducer) prior to the depletion of sulfate in the medium, but this lag phase was not observed in follow on experiments (see3.7. Effect of particle size section). The lag phase may also represent a transition between yeast extract utilization and coal utilization.

Effect of pH

In this experiment, the biological production of methane from Wyodak coal was evaluated as a function of medium pH. The media were prepared at pH levels of 6.1, 7.6, and 8.3, which correspond to the pKa values of the Good buffers that were added (2.0 g/L MES, TES, and Bicine, respectively). The medium preparation protocol was modified so that pH adjustment was done on an individual tube basis after the addition of sodium bicarbonate and coal. Following autoclave sterilization in the presence of coal, pH levels at the time of inoculation were 6.4, 7.4, and 7.6, respectively. As shown in Fig. 5, cultures grown at a pH of 6.4 had substantially higher methane production levels than those grown at either 7.4 or 7.6. The maximum methane production rate for pH 6.4 was 21 μmol/g coal/day (between 216 and 264 h), compared to 2.7 and 4.4 μmol/g coal/day (between 432 and 624 h) at pH 7.4 and 7.6, respectively. Methane production did not vary with pH during the first 72 h; again, this methane level is consistent with the rapid conversion of yeast extract present in the medium (all tubes were inoculated from a pH 7.3 culture). At the final time of 624 h, the total methane present in the pH 6.4 cultures (0.197 mmol/g coal) was over 200% greater than methane levels in the pH 7.4 or 7.6 cultures (0.059 and 0.064 mmol/g coal, respectively). Final pH values were measured in the anaerobic chamber at 624 h; average values were 6.4, 7.6, and 7.8 for cultures with an initial pH of 6.4, 7.4, and 7.6, respectively. No detectable methane was observed for killed cell controls.

Effect of Particle Size

This experiment was designed to evaluate the effect of coal particle size on the rate and extent of biological methane production.

Three coal particle size ranges were used: 80–140 mesh (106 to 180 μm), 30–60 mesh (250 to 600 μm), and 12–20 mesh (850 to 1700 μm). As shown in Fig. 6, the small particle cultures had higher methane production rates than either the medium or large particle cultures. The initial production rate (0 to 120 h) in the small particle cultures (5.1 μmol/g coal/day) was 19% higher than that of medium particle cultures (4.3 μmol/g coal/day) and 34% higher than that of large particle cultures (3.8 μmol/g coal/day). The small particle cultures had a maximum methane production rate of 6.6 μmol/g coal/day, which was 113% greater than the maximum rate for medium particles (3.1 μmol/g coal/day), and 214% greater than the maximum rate for large particles (2.1 μmol/g coal/day). No detectable methane was observed for killed cell controls. A plot of the maximum methane production rate versus external coal particle surface area exhibited a linear trend with a non-zero intercept (r^2 = 0.9991, Fig. 7), which is consistent with a mass transfer limited process. The non-zero intercept on this plot is consistent with the simultaneous conversion of some yeast extract to methane.

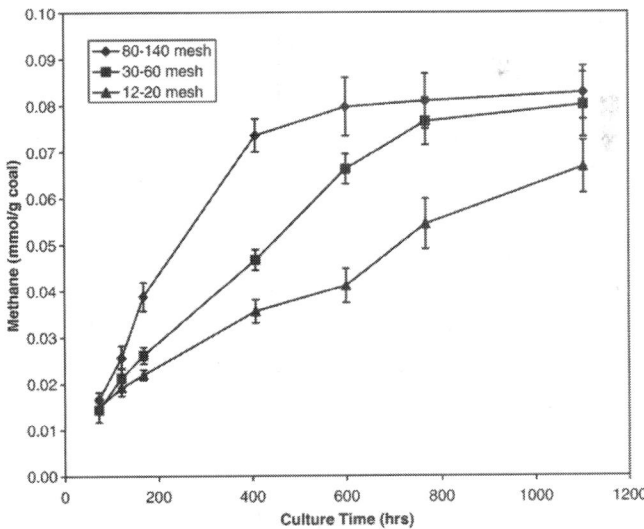

Figure 6: Microbial methane production from Wyodak coal at 30 °C, pH 7.3 for 12–20, 30–60, and 80–140 mesh coal particle size fractions. Error

bars represent 1 standard deviation for triplicate cultures. No methane was observed in killed cell controls.

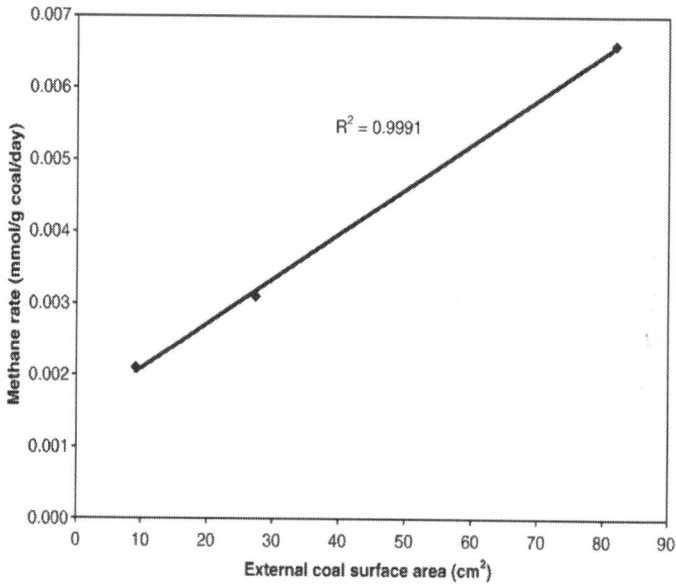

Figure 7: Maximum methane production rates from Fig. 5, plotted as a function of the external coal surface area for that particle size fraction. External surface areas were estimated from the mean diameter for each particle size fraction (1260, 422, and 141 μm for mesh sizes 12–20, 30–60, and 80–140, respectively) assuming uniform spherical particles with a specific gravity of 1.3.

Methane levels at 72 h were once again independent of the variable parameter (particle size in this case) and near the expected yield for rapid conversion of yeast extract. Unlike the temperature and pH experiments, no lag phase was observed after 72 h for any of the particle size ranges tested; this experiment was performed after the temperature and pH experiments, and the maintenance culture inoculum may have been better adapted to coal as the primary carbon source at this point. It should be noted that the final plateau in methane production for the small particle tubes began earlier than for other particle sizes; this may have been caused by a yield limiting nutrient other than coal, as rapid methane production in the small particle cultures would have caused this yield limiting nutrient to be exhausted earlier than in the other cultures. At the final sample time of 1100 h the medium particle cultures had approached the same methane level as the small particle cultures. There are four factors that could lead to a halt in methane production in these closed batch cultures: (1) depletion of nutrients supplied by the coal solids, (2) depletion of a non-coal nutrient(s) in the aqueous medium, (3) accumulation of toxic aqueous products, and/or (4) accumulation of methane in the closed headspace. To determine the effect of methane accumulation on methane production, the gas headspace of two of the small particle tubes was completely replaced with pure nitrogen at 300 kPa. Apparently methane had not accumulated to inhibitory levels, as no new methane production was observed after the headspace exchange. Given that only 0.13 wt.% of the coal was converted to methane (based on final methane levels for the 80–140 mesh and 30–60 mesh cultures), it seems unlikely that coal was the yield limiting nutrient in these experiments.

Effect of Solvent Addition

The goal of this experiment was to determine if the addition of water-miscible solvents at low concentrations would enhance the microbial conversion of Wyodak coal to methane. It was hypothesized that miscible organic solvents would enhance the solubility of key coal substrates, which would also enhance the rate of mass transfer into the aqueous medium. Coal cultures were supplemented with methanol, N,N-dimethylformamide (DMF), and pyridine. These solvents were chosen for their biocompatibility as well as for their ability to solvate coal compounds. As mentioned in Section 3.1 Substrate and nutrient results, this consortium tested positive for methane production from 0.5 vol.% methanol. As E. coli is able to grow in the presence of 5 vol.% methanol, and will tolerate methanol concentrations up to 15 vol.% (Coryell, 2003), it was decided to test a higher concentration of methanol (2 vol.%) in the current experiment. Pyridine is a conventional coal solvent that has been successfully used in the biological solubilization of coal using dissolved enzymes (Scott and Lewis, 1988). Other studies have shown that pyridine can also act as a primary carbon substrate in energy metabolism (Sims et al., 1986 and Rhee et al., 1997). Pyridine was added to cultures at 0.1 vol.%, which lies within the concentration range given (0.025 to 0.177 vol.%) for the anaerobic degradation of pyridine by denitrifying bacteria without inhibitory effects (Rhee et al., 1997). N,N-dimethylformamide (DMF) is a widely used industrial solvent. This compound is not commonly degraded by microorganisms, and has been shown to be inhibitory to many species; however, a species able to utilize DMF as a sole source of carbon and nitrogen has been isolated from soil samples associated with coal tailings (Veeranagouda et al., 2006). Further studies have shown that some E. coli strains can grow on agar plates treated with DMF, and that single mutations can greatly increase DMF resistance (Selifonova et al., 2001). Veeranagouda et al. (2004) describe the growth of bacteria on DMF at concentrations up to 0.4 vol.% a reduced concentration of DMF (0.25 vol.%) was used in this experiment because of its known toxicity.

Fig. 8 shows biological methane production from Wyodak coal cultures in the presence of methanol, pyridine, and DMF, with results for a no-solvent control culture provided for comparison. Methanol at 2 vol.% was inhibitory, as these cultures exhibited no detectable methane production over 1140 h of growth time. One of the cultures supplemented with 0.1 vol.% pyridine showed no methane production over 1100 h, and was excluded from the data shown in Fig. 8. Judging from the remaining pyridine culture, 0.1 vol.% pyridine had no effect on methane production, as the methane levels were equal to those observed in the no-solvent control culture throughout the experiment. The addition of DMF to the coal cultures had a significant positive effect on methane production; at 1100 h, the cultures treated with 0.25 vol.% DMF had produced 346% more methane than the no-solvent control cultures. No detectable methane was observed for killed cell controls.

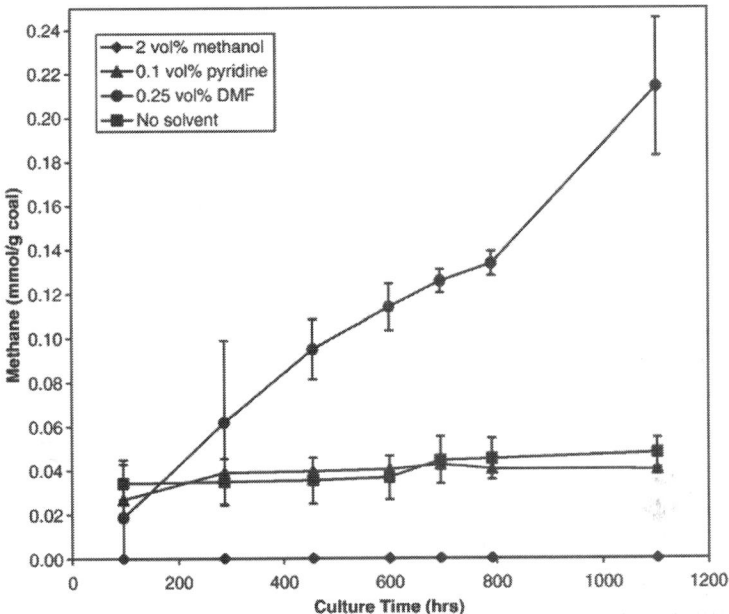

Figure 8: The effect of methanol (2 vol.%), pyridine (0.1 vol.%), and DMF (0.25 vol.%) solvent addition on microbial methane production from 0.25 g of 30–60 mesh Wyodak coal at 30 °C, pH 7.3. Error bars represent 1 standard deviation for duplicate cultures. Pyridine data represents 1 culture only. No methane was observed in no-coal or killed cell controls.

Methane production for all coal cultures was similar over the first sampling period of 96 h, except for the 2 vol.% methanol cultures which showed no methane production. Since all three solvents have been shown to be substrates for other microorganisms, coal-free control cultures were also prepared with the same concentrations of each solvent. No methane was detected for any of these cultures over the 1140 h growth time, suggesting that none of these solvents acted as a substrate for the PRB consortium at the concentrations tested; however, these controls were not conclusive, as some methane should have been produced from the yeast extract present in the medium. It is possible that in a coal-free system, pyridine/ DMF are also inhibitory at the concentrations tested, but solvent sorption allowed methanogenesis to occur in the presence of coal.

DISCUSSION

Significance of Observed 16S rDNA Sequences

These clone libraries represent the first report of 16S rDNA sequences cloned from a methanogenic culture enriched from a coalbed methane well. Each OTU generated in this study represents a phylotype and may be representative of a bacterial species. Some of the low-similarity phylotypes may represent novel species specifically adapted to a coalbed ecosystem. For distantly related phylotypes, the extrapolation of functional properties from well-characterized cultured strains may not be justified. However, the phylogenetic similarity of some clones to cultured species with known physiology and substrate spectra, i.e. fermentative (*Clostridium* sp., *Acholeplasma* sp., and *Spirochaeta* sp.), acetogenic (*A.hydrogenoformans* and *Syntrophomonas* sp.) and acetoclastic methanogen (*Methanosarcina* sp.) species, is consistent with the different trophic groups required to convert complex organic matter to methane. The limited archaeal diversity of our consortium is not surprising given the nutrient-limited environment of a coal seam; for an Antrim Shale well water, only one phylotype related to *Methanocorpusculum bavaricum* was found in the archaeal clone library (Waldron et al., 2007).

Whiticar et al. (1986) report that acetate fermentation is the dominant process by which methane is produced in freshwater. There are only two genera of methanogens that are capable of using acetate for methanogenesis: *Methanosarcina* sp. and *Methanosaeta* sp. (Zinder, 1993). The similarity of our two archaeal OTUs to *M. mazei* SarPi is consistent with the observation that the culture uses acetate and methanol rather than H_2–CO_2 for methane production. Methanogenic substrates for *M. mazei* include acetate and methylated C1 compounds; H_2–CO_2 is used very slowly or not at all (Mah and Kuhn, 1984). This is in contrast to other tests of

Powder River Basin coalbed consortia, which indicated that the methanogens present could utilize both acetate and H_2–CO_2 as substrates (LUCA Technologies, 2004). Based on values for methane $\delta^{13}C$ and δD isotopes, Faiz and Hendry (2006) state that Powder River Basin gases appear to be derived from acetoclastic reactions. However, Gorody (1999) uses $\delta^{13}C_{CO2}$ values and fractionation factors to conclude that PRB methane is the product of biological CO_2 reduction (Flores et al., 2008-this volume). It is possible that the methane in place was produced via a pathway that is different from the pathway(s) that are currently active in the coal seam. It is also possible that CO_2-reducing methanogens present in the well water were lost during the sampling/enrichment process; some oxygen exposure may have occurred during field transfers, or inhibitory compounds formed during autoclave sterilization of our coal slurry medium may have been present.

Acetogens play a critical role in the biological conversion of complex substrates (like coal) to methane. These organisms reduce intermediate metabolites into the acetate/H_2–CO_2 substrates required of methanogens (McInerney and Bryant, 1981). The acetogen A. hydrogenoformans was first isolated from estuarine mud with glutamate as the substrate (Stams and Hansen, 1984). It utilizes α-ketoglutarate, adenine, pyruvate, oxaloacetate, citrate, and a number of amino acids as substrates, but not sugars. When grown on pyruvate in pure culture, acetate was the major fermentation product. Isbister and Barik (1993)report pyruvate has been identified as an intermediate in the biological production of methane from coal; this suggests that OTUs similar to A. hydrogenoformans may play an important role as acetogens converting coal intermediates into the acetate needed by Methanosarcina sp. Degradation of certain substrates by A. hydrogenoformans was enhanced by or dependent upon co-cultivation with a H_2-utilizing partner (Stams and Hansen, 1984), suggesting that an H_2-utilizing methanogen may be present in-situ but was not recovered in our sampling-enrichment method.

Temperature and pH Effects

The higher methane production rates seen in this study at higher temperatures are in agreement with results reported by other groups. Zeikus and Winfrey (1976) reported an optimum temperature range for biogenic gas production of 35 to 42 °C for methanogens taken from a freshwater lake, which was considerably higher than the maximum observed lake temperature of 23 °C. For our results, the observed optimum of 38 °C was higher than the sampled reservoir temperature of 17.5 °C. Harding et al. (1993) investigated the production of methane from a Texas lignite coal using a termite gut consortium at 28, 37, and 40 °C, and found that methane production increased with increasing temperature. The optimum temperature for *M. mazei* is reported as between 35 and 42 °C (Boone et al., 1993), while *A. hydrogenoformans* has a temperature range of 15 to 42 °C with an optimum of 30 °C (Stams and Hansen, 1984). Enhanced cell metabolism and growth kinetics may not be the only reason for higher methanogenesis rates; the aqueous solubility of coal substrates will also increase with temperature. This in turn will increase the rate and extent of substrate mass transfer from the coal solids. If dissolution represents the rate-limiting step in methane production from coal solids, increased solubility will lead to enhanced methanogenesis.

While pH optima for methanogenesis are typically near neutrality, acidophilic species have been cultured that are able to produce methane down to pH 3.0 (Williams and Crawford, 1985). *M. mazei* normally has a pH optimum between 6 and 7 (Boone et al., 1993). While acetogens tend to prefer a more acidic environment (Isbister and Barik, 1993), 6.7 was reported as the lower pH limit for growth of *A. hydrogenoformans* on glutamate (Stams and Hansen, 1984). Volkwein et al. (1994) evaluated methane production from coal for six different consortia on three different coals (high volatile A bituminous, subbituminous C, and low volatile bituminous in rank) at pH 5 and 7. For 12 of the 18 consortium-coal combinations evaluated, methane production was greater at the lower pH. Alternatively, in the bioconversion of a Texas lignite coal using a termite gut consortium, methane production did not vary from pH 6 to 7 (Harding et al., 1993). The increased production level seen in our pH 6.4 cultures may have been caused by an enhanced growth rate of the PRB consortium at this condition. Also, an acidic pH may enhance methane production rates by enhancing coal solubility; acids may enter the coal pore structure and interact with ion-exchangeable cations, resulting in limited dissolution of the coal via disruption of ionic bridges (Faison, 1992). Acids may also hydrolyze ester or ether bonds within the coal matrix (Faison, 1992). While the pH of the coal surface may differ from the bulk medium pH because of charged surface groups like carboxylic acids (Faison, 1992), no microbes were apparent on the surface of coal harvested from a maintenance culture; the 100 rpm agitation level may have prevented the formation of surface biofilms in our cultures.

Evidence of Mass Transfer Limitations

While initial methane production in these cultures (Figs. 4–6) may be due to the conversion of yeast extract, stoichiometry calculations show that methane accumulations greater than 0.021 mmol/g coal represent the conversion of Wyodak coal to methane. The increase in methane production rates in direct proportion to coal surface area serves as evidence that this system was mass transfer limited, as mass transfer rates are directly proportional to interfacial surface area (Gilcrease et al., 2001). Scott (1999) hypothesizes that the amount of biologically generated methane will be proportional to the cleat surface area of the coalbed, and uses coal surface area to extrapolate laboratory rates to subsurface rates. While Harding et al. (1993)found no significant correlation between coal particle size and the rate of biological methane production, they were using a Texas lignite, which is most likely more soluble than the subbituminous B Wyodak coal used in this study. As coal solubility increases/microbial rates decrease, a given system can shift from one that is mass transfer limited to one limited by the metabolic rate of the microbial consortium. It should be noted that the sieve classification of our Wyodak coal could have concentrated different macerals into different size fractions, which could also affect the rate of methane production. It has been shown that different coal macerals will biodegrade at different rates (Isbister and Barik, 1993).

While not conclusive, enhanced methane production rates from coal with the addition of 0.25 vol.% DMF solvent is also consistent with a mass transfer limited system. Solvents can enhance the aqueous solubility of coal compounds, which increases the concentration driving force for mass transfer between the coal solids and the aqueous medium. If DMF did not serve as a substrate for methane production, the enhanced methane production rates were likely due to higher solid–liquid mass transfer rates. It is also possible that DMF enhanced substrate transport through the cell wall/membrane by increasing cell permeability (Leon et al., 1998). In the case of 2 vol.% methanol, cell permeability may have been increased to the point that cell death occurred. The fact that no methane was produced in the no-coal controls was unexpected, as some methane should have been produced from the 50 mg/L yeast extract present in the medium; thus, we cannot conclude that DMF does not provide an additional carbon source for biological methane production. Given that the methanogens present in this consortium are known methylotrophs and that the chemical structure of DMF includes a dimethylamine group, it seems feasible that DMF could be used directly as a methanogenic substrate; the estimated methane yield from the 0.25 vol.% DMF present in this culture is ≥ 1.3 mmol/g coal. As such, we cannot conclude whether enhanced methane production in the presence of DMF was due to enhanced coal utilization or the direct utilization of DMF.

Implications for Microbially Enhanced Coalbed Methane (MECoM)

This study has demonstrated that well-bore water samples collected from a PRB coalbed methane well contain a microbial consortium capable of converting Wyodak coal to methane in a controlled laboratory environment. This suggests that a viable methanogenic consortium is still present in PRB coal seams, and that the *in-situ* bioconversion of coal to methane could be stimulated with the proper manipulation of the coal seam environment. The culture with the highest methane productivity was the pH 6.4 culture (Fig. 5); the maximum methane production rate was 21 µmol/g coal/ day, which equals 0.50 m^3/t coal/day (16 scf/ton/day). The overall average rate for this culture was 7.6 µmol/g coal/day or 0.18 m^3/t coal/day (5.8 scf/ton/day). An estimate of the overall gas content of the Powder River Basin is 1.6 m^3/t coal (50 scf/ton) (De Bruin et al., 2000); *if* these laboratory conditions could somehow be replicated *in-situ*, new biological gas generation could replenish existing reserves in less than 10 days. A more realistic estimate for enhanced *in-situ* methane production potential from a 17.5 °C, pH 8 Wyodak coal seam would be to take rates from the lowest temperature (22 °C, pH 7.3) experiment and extrapolate based on coal surface area. Using an estimated *in-situ* coal surface area of 351 m^2/t coal (Scott, 1999), a 22 °C Wyodak seam supplemented with the nutrients provided in our medium could produce methane at a rate of 0.0030 m^3/t coal/day (0.094 scf/ton/day). At this rate it would take 533 days to produce 1.6 m^3 of methane per metric ton of coal, which is still significant from a production standpoint. It should be noted that air exposure during sieving and/or the autoclaving of our Wyodak coal may have increased its bioavailability; as such, extrapolation of our laboratory methanogenesis rates may overestimate potential *in-situ* rates.

It is also interesting to compare methane production rates from this study with other coal biogasification laboratory studies. A coal-formation water sample from the Dietz coalbed in northern Sheridan County, WY (see Fig. 1 for county location and Flores et al., 2008-this volume for stratigraphic position) supplemented with an undisclosed amendment produced 0.282 m^3 per metric ton of coal over 159 days (LUCA Technologies, 2004). This corresponds to an overall average rate of 0.0018 m^3/t coal/day (0.056 scf/ton/day), which is less than observed in our 22 °C laboratory culture (0.057 m^3/t coal/day), but comparable to the 22 °C *in-situ* rate predicted above. Incubation temperature and coal particle size could explain why the Dietz rates were lower, but these values were not reported. In a separate study, a methanogenic consortium derived from a wood-eating termite was used to convert a Texas lignite coal to methane at 37 °C (Harding et al., 1993); after correcting for methane produced from non-coal nutrients, maximum methane production rates of 12 to 17 m^3/t coal/day were observed for coal particle sizes ranging from 28 to 325 mesh. These rates are significantly higher than any of the rates observed in this study; either the termite-derived consortium is more robust than our PRB consortium, and/or the solubility/bioavailability of the Texas lignite is higher than that of Wyodak subbituminous coal. Another possibility is that the overall cell concentrations may have been much higher compared to our cultures. It is not known whether a such a foreign microbial consortium could thrive and produce methane if introduced to a coal seam environment.

For the *in-situ* stimulation of methanogenic consortia in coal beds, the addition of non-carbon/energy nutrients will be important, and will most likely be specific for the type of coal and consortium present at that site. Assuming that coal is not the yield limiting nutrient, new *in-situ* methane production could be sustained for a longer period of time with the continuous pumping of supplemental nutrients. However, if microbial activity is enhanced enough, at some point the coal solid-microbe system will become mass transfer limited (Gilcrease et al., 1996). The results of this study indicate that mass transfer can be limiting, and may be affected by a number of environmental parameters. These parameters could be used to screen for promising MECoM sites, and/or used to suggest potential stimulation treatments. Increasing coalbed temperatures *in-situ* may not be economically feasible (then again, steam flooding is used in enhanced oil recovery operations), but temperature may be an important selection criterion for MECoM sites due to its effect on metabolic activity and coal solubility. Similarly, it has been proposed that surface area be used as a means of estimating MECoM potential in a particular coal seam (Scott, 1999), and *in-situ* fracing operations could be used to enhance coal surface area. If pH does have a positive effect on coal solubility, the addition of acid may be more feasible (from an economic and environmental standpoint) than the addition of solvents; previous experience with the injection of organic solvents resulted in lowered coalbed permeability and reservoir damage rather than enhanced methane producibility (Scott, 1999).

ACKNOWLEDGMENTS

This work was supported in part by a graduate fellowship from NSF EPSCoR. We thank Welldog and Peabody Natural Gas for providing the well sample, and Dr. Ralph Tanner, University of Oklahoma, for his instruction in anaerobic techniques. Drs. Rajesh Sani and Gurdeep Rastogi, South Dakota School of Mines and Technology, generously provided the phylogenetic analysis. Finally, we thank Khang Vo for his contributions in the laboratory.

REFERENCES

1. Altschul, S.F., Gish, W., Miller, W., Myers, E.W., Lipman, D.J., 1990. Basic local alignment search tool. Journal of Molecular Biology 215, 403–410.

2. Amann, R.I., Ludwig, W., Schleifer, K.H., 1995. Phylogenetic identification and in situ detection of individual microbial cells without cultivation. Microbiological Review 59, 143–169.

3. Bailey, J.E., Ollis, D.F., 1986. Biochemical Engineering Fundamentals, Second Ed. McGraw-Hill, New York.

4. Baker, K.H., 1994. Bioremediation of surface and subsurface soils. In: Baker, K.H., Herson, D.S. (Eds.), Bioremediation. McGraw-Hill, New York, pp. 203–259.

5. Balch, W.E., Wolfe, R.S., 1976. New approach to the cultivation of methanogenic bacteria: 2-mercaptoethanesulfonic acid (HS-CoM)-dependent growth of Methanobacterium

ruminantium in a pressurized atmosphere. Applied Environmental Microbiology 32 (6), 781–791.

6.   Boone, D.R., Whitman, W.B., Rouviere, P., 1993. Diversity and taxonomy of methanogens. In: Ferry, J.G. (Ed.), Methanogenesis: Ecology, Physiology, Biochemistry, and Genetics. Chapman and Hall, New York, pp. 35–80.

7.   Borneman, J., Skroch, P.W., O'Sullivan, K.M., Palus, J.A., Rumjanek, N.G., Jansen, J.L., Nienhuis, J., Triplett, E.W., 1996. Molecular microbial diversity of an agricultural soil in Wisconsin. Applied Environmental Microbiology 62 (6), 1935–1943.

8.   Bräuer, S.L., Yashiro, E., Ueno, N.G., Yavitt, J.B., Zinder, S.H., 2006. Characterization of acid-tolerant H2/CO2-utilizing methanogenic enrichment cultures from an acidic peat bog in New York State. FEMS (Federation of European Microbiological Societies) Microbiological Ecology 5, 206–216.

9.   Colberg, P.J., Young, L.Y., 1985. Aromatic and volatile acid intermediates observed during anaerobic metabolism of lignin-derived oligomers. Applied Environmental Microbiology 49 (2), 350–358.

10.   Cookson Jr., J.T., 1995. Bioremediation Engineering Design and Application. McGrawHill, New York.

11.   Coryell, G.T., II, 2003. The production of hydroxylated diphenyl compounds using whole-cell biocatalysis. M. S. Thesis, University of Wyoming, Laramie, WY.

12.   Curtis, T.P., Head, I.M., Lunn, M., Woodcock, S., Schloss, P.D., Sloan, W.T., 2006. What is the extent of prokaryotic diversity? Philosophical Transactions Royal Society of London Series B Biological Sciences 361 (1475), 2023–2037.

13.   De Bruin, R.H., Lyman, R.M., Jones, R.W., Cook, L.W., 2000. Coalbed Methane in Wyoming, Wyoming State Geological Survey, Information Pamphlet 7, Laramie, WY.

14.   Faison, B.D., 1992. The chemistry of low rank coal and its relationship to the biochemical mechanisms of coal

biotransformation. In: Crawford, D.L. (Ed.), Microbial Transformations of Low Rank Coals. CRC (Chemical Rubber Company) Press, Boca Raton, FL, pp. 1–26.

15. Faiz, M., Hendry, P., 2006. Significance of microbial activity in Australian coal bed methane reservoirs—a review. Bulletin of Canadian Petroleum Geology 54 (3), 261–272.

16. Fakoussa, R.M., Hofrichter, M., 1999. Biotechnology and microbiology of coal degradation. Applied Microbiological Biotechnology 52, 25–40.

17. Fellows Energy, 2005. The case for coalbed methane gas. http://www.fellowsenergy. com/index.php?id=85.

18. Felsenstein, J., 1989. PHYLIP—Phylogeny Inference Package (Version 3.2). Cladistics 5, 164–166.

19. Flanegan, K.C., 2004. Characterization of methanogenic consortia in Powder River Basin coalbed methane wells. M. S. Thesis, South Dakota School of Mines and Technology, Rapid City, SD.

20. Flores, R.M., 1999. Wyodak-Anderson coal zone in the Powder River basin, Wyoming and Montana: a tale of uncorrelatable coal beds. In: Miller, W.R. (Ed.), Fiftieth Field Conference Guidebook—Coalbed Methane and the Tertiary Geology of the Powder River Basin Wyoming and Montana. Wyoming Geological Association, Casper, WY, pp. 1–24.

21. Flores, R.M., 2004. Coalbed Methane in the Powder River Basin, Wyoming and Montana: An Assessment of the Tertiary–Upper Cretaceous Coalbed Methane Total Petroleum System. U.S. Geological Survey Digital Data Series DDS-69-C.

22. Flores, R.M., Rice, C.A., Stricker, G.D., Warden, A., Ellis, M.S. 2008. Methanogenic pathways of coalbed gas in the Powder River Basin, United States: The geologic factor. In: Flores, R.M. (Ed.), Microbes, Methanogenesis, and Microbial Gas in Coal. International Journal of Coal Geology Special Issue, 76, 52–75 (this volume).

23. Garrity, G.M. (Ed.), 2005. Bergey's Manual of Systematic Bacteriology, second edition. Springer, New York. 2.

24. Gilcrease, P.C., 1997. Mass transfer effects on the bioreduction of TNT solids in slurry reactors. PhD dissertation, Colorado State University, Fort Collins, CO.

25. Gilcrease, P.C., Murphy, V.G., Reardon, K.F., 1996. Bioremediation of solid TNT particles in a soil slurry reactor: mass transfer considerations. In: Erickson, L.E., Tillison, D.L.,

26. Grant, S.C., McDonald, J.P. (Eds.), HSRC/WERC Joint Conference on the Environment.

27. Great Plains/Rocky Mountain Hazardous Substance Research Center, Albuquerque, NM, pp. 152–162.

28. Gilcrease, P.C., Murphy, V.G., Reardon, K.F., 2001. Simultaneous grinding and dissolution of TNT solids in an agitated slurry. AIChE (American Institute of Chemical Engineers) Journal 47 (3), 572–581.

29. Godon, J.-J., Zumstein, E., Dabert, P., Habouzit, F., Moletta, R., 1997. Molecular microbial diversity of an anaerobic digestor as determined by small-subunit rDNA sequence analysis. Applied Environmental Microbiology 63 (7), 2802–2813.

30. Gorody, A.W., 1999. The origin of natural gas in the tertiary coal seams on the eastern margin of the Powder River Basin. In: Miller, W.R. (Ed.), Coalbed Methane and Tertiary Geology of the Powder River Basin: Wyoming Geological Association

31. Guidebook, 50th Annual Field Conference, pp. 89–101.

32. Grabowski, A., Nercessian, O., Fayolle, F., Blanchet, D., Jeanthon, C., 2005. Microbial diversity in production waters of a low-temperature biodegraded oil reservoir. FEMS (Federation of European Microbiological Societies) Microbiological Ecology 54 (3), 427–443.

33. Harding, R., Czarnecki, S., Isbister, J., Barik, S., 1993. Biogasification of low-rank coal. TR- 101572, ARCTECH, Inc. for Electric Power Research Institute, Chantilly, VA.

34. Hungate, R.E., 1969. A roll tube method for cultivation of strict anaerobes. In: Norris, J.R., Ribbons, D.W. (Eds.), Methods in Microbiology. Academic Press Inc., New York, pp. 117–132.

35. Isbister, J.D., Barik, S., 1993. Biogasification of low rank coals. In: Crawford, D.L. (Ed.), Microbial Transformations of Low Rank Coals. CRC (Chemical Rubber Company) Press, Boca Raton, FL, pp. 139–156.

36. Joulian, C., Ollivier, B., Patel, B.K.C., Roger, P.A., 1998. Phenotypic and phylogenetic characterization of dominant culturable methanogens isolated from ricefield soils. FEMS (Federation of European Microbiological Societies) Microbiology Ecology 25 (2), 135–145.

37. Kemp, P.F., Aller, J.Y., 2004. Bacterial diversity in aquatic and other environments: what 16S rDNA libraries can tell us. FEMS (Federation of European Microbiological Societies) Microbiological Ecology 47 (2), 161–177.

38. Kumar, S., Tamura, K., Nei, M., 1993. MEGA: Molecular Evolutionary Genetics Analysis. Pennsylvania State University, University Park, PA.

39. Leclerc, M., Delgenes, J.P., Godon, J.J., 2004. Diversity of the archaeal community in 44 anaerobic digesters as determined by single strand conformation polymorphism analysis and 16S rDNA sequencing. Environmental Microbiology 8, 809–819.

40. Leon, R., Fernandes, P., Pinheiro, H.M., Cabral, J.M.S., 1998. Whole-cell biocatalysis in organic media. Enzyme and Microbial Technology 23, 483–500.

41. LUCA Technologies, LLC, 2004. Active biogenesis of methane in Wyoming's Powder River Basin. http://www.lucatechnologies.com/content/index.cfm? fuseaction=showContent&contentID=22&navID=22.

42. Mah, R.A., Kuhn, D.A., 1984. Transfer of the type species of the genus Methanococcus to the genus Methanosarcina, naming it Methanosarcina mazei (Barker 1936) comb. nov. et emend. and conservation of the genus Methanococcus (approved lists 1980) with Methanococcus vannielii (approved lists 1980) as the type species. International Journal of Systemic Bacteriology 34 (2), 263–265.

43. Maidak, B.L., Olsen, G.J., Larsen, N., Overbeek, R., McCaughey, M.J., Woese, C.R., 1997. The RDP (ribosomal database project). Nucleic Acids Research 25, 109–110.

44. Mavor, M., Pratt, T., DeBruyn, R., 1999. Study quantifies Powder River coal seam properties. Oil & Gas Journal, April 26, 1999, 35–40.

45. McInerney, M.J., Bryant, M.P., 1981. Review of methane fermentation fundamentals. In: Wise, D.L. (Ed.), Fuel Gas Production from Biomass. CRC (Chemical Rubber Company) Press, Boca Raton, FL, pp. 19–46.

46. Meijer, W.G., Nienhuis-Kuiper, M.E., Hansen, T.A., 1999. Fermentative bacteria from estuarine mud: phylogenetic position of Acidaminobacter hydrogenoformans and description of a new type of Gram-negative, propionigenic bacterium as Propionibacter pelophilus gen. nov., sp. nov. International Journal of Systemic Bacteriology 49, 1039–1044.

47. Ogram, A.V., Jessup, R.E., Ou, L.T., Rao, P.S.C., 1985. Effects of sorption on biological degradation rates of (2,4-dichlorophenoxy)acetic acid in soils. Applied Environmental Microbiology 49 (3), 582–587.

48. Prescott, L.M., Harley, J.P., Klein, D.A., 1990. Microbiology. William C. Brown, Dubuque, IA.

49. Ramaswami, A., Luthy, R.G., 1997. Measuring and modeling physicochemical limitations to bioavailability and biodegradation. In: Hurst, C.J. (Ed.), Manual of Environmental

50. Microbiology. American Society for Microbiology, pp. 721–729. Rhee, S.-K., Lee, G.M., Yoon, J.-H., Park, Y.-H., Bae, H.-S., Lee, S.-T., 1997. Anaerobic and aerobic degradation of pyridine by a newly isolated denitrifying bacterium. Applied Environmental Microbiology 63 (7), 2578–2585.

51. Rice, D.D., 1993. Composition and origins of coalbed gas. In: Law, B.E., Rice, D.D. (Eds.), Hydrocarbons from Coal, Studies in Geology #38. American Association of Petroleum Geologists, Tulsa, OK, pp. 159–184.

52. Rossetti, S., Blackall, L.L., Majone, M., Hugenholtz, P., Plumb, J.J., Tando, V., 2003. Kinetic and phylogenetic characterization of an anaerobic dechlorinating microbial community. Microbiology 149, 459–469.

53. Saitou, N., Nei, M., 1987. The neighbor-joining method: a new method for reconstructing phylogenetic trees. Molecular Biological Evolution 4, 406–425.

54. Schloss, P.D., Handelsman, J., 2005. Introducing DOTUR, a computer program for defining operational taxonomic units and estimating species richness. Applied Environmental Microbiology 71, 1501–1506.

55. Schloss, P.D., Handelsman, J., 2006. Toward a census of bacteria in soil. PLoS Computational Biology 2 (7), e92.

56. Scott, A.R., 1999. Improving coal gas recovery with microbially enhanced coalbed methane. In: Mastalerz, M., Glikson, M., Golding, S.D. (Eds.), Coalbed Methane:

57. Scientific, Environmental and Economic Evaluation. Kluwer, Dordrecht, pp. 89–110.

58. Scott, A.R., Kaiser, W.R., Ayers Jr., W.B., 1994. Thermogenic and secondary biogenic gases, San Juan basin, Colorado and New Mexico—implications for coalbed gas producibility. American Association of Petroleum Geologists Bulletin 78 (8), 1186–1209.

59. Scott, C.D., Lewis, S.N., 1988. Biological solubilization of coal using both in vivo and in vitro processes. Applied Biochemistry and Biotechnology 18, 403–412.

60. Sekiguchi, Y., Kamagata, Y., Syutsubo, K., Ohashi, A., Harada, H., Nakamura, K., 1998. Phylogenetic diversity of mesophilic and thermophilic granular sludges determined by 16S rRNA gene analysis. Microbiology 144, 2655–2665.

61. Selifonova, O., Valle, F., Schellenberger, V., 2001. Rapid evolution of novel traits in microorganisms. Applied Environmental Microbiology 67 (8), 3645–3649.

62. Sims, G.K., Sommers, L.E., Konopka, A., 1986. Degradation of pyridine by Micrococcus luteus isolated from soil. Applied Environmental Microbiology 51 (5), 963–968.

63. Skladany, G.J., Baker, K.H., 1994. Laboratory biotreatability studies. In: Baker, K.H., Herson, D.S. (Eds.), Bioremediation. McGraw-Hill, New York, pp. 97–172.

64. Smith, J.M., Van Ness, H.C., Abbott, M.M., 2005. Introduction to Chemical Engineering Thermodynamics, Seventh Ed. McGraw-Hill, New York.

65. Stams, A.J.M., Hansen, T.A., 1984. Fermentation of glutamate and other compounds by Acidaminobacter hydrogenoformans gen. nov. sp. nov., an obligate anaerobe isolated from black mud. Studies with pure cultures and mixed cultures with sulfatereducing and methanogenic bacteria. Archives of Microbiology 137 (4), 329–337.

66. Swindell, G.S., 2007. Powder River basin coalbed methane wells—reserves and rates. Presented at the 2007 SPE Rocky Mountain Oil & Gas Technology Symposium, Denver, CO. April 16–18.

67. Tanner, R.S., 2002. Cultivation of bacteria and fungi, In: Hurst, C.J., Crawford, R.L., Knudsen, G.R., McInerney, M.J., Stetzenbach, L.D. (Eds.), Manual of Environmental Microbiology, second edition. ASM (American Society for Microbiology) Press, Washington, D.C., pp. 62–70. Thompson, J.D., Higgins, D.G., Gibson, T.J., 1994. CLUSTALW: improving the sensitivity of progressive multiple sequence alignment through sequence weighting, positionspecific gap penalties and weight matrix choice. Nucleic Acids Research 22, 4673–4680.

68. Veeranagouda, Y., Emmanuel Paul, P.V., Gorla, P., Siddavattam, D., Karegoudar, T.B., 2006. Complete mineralisation of dimethylformamide by Ochrobactrum sp. DGVK1 isolated from the soil samples collected from the coalmine leftovers. Applied Microbiology Biotechnology 71 (3), 369–375.

69. Veeranagouda, Y., Patil, K.N., Karegoudar, T.B., 2004. A method for screening of bacteria capable of degrading dimethylformamide. Current Science 87 (12), 1652–1654.

70. Volkwein, J.C., Schoeneman, A.L., Clausen, E.G., Gaddy, J.L., Johnson, E.R., Basu, R., Ju, N., Klasson, K.T., 1994. Biological production of methane from bituminous coal. Fuel Processing Technology 40, 339–345.

71. Waldron, P.J., Petsch, S.T., Martini, A.M., Nüsslein, K., 2007. Salinity constraints on subsurface archaeal diversity and methanogenesis in sedimentary rock rich in organic matter. Applied Environmental Microbiology 73 (13), 4171–4179.

72. Wani, A.A., Surakasi, V.P., Siddharth, J., Raghavan, R.G., Patole, M.S., Ranade, D., Shouche,

73. Y.S., 2006. Molecular analyses of microbial diversity associated with the Lonar soda lake in India: an impact crater in a basalt area. Research In Microbiology 57, 928–937.

74. Whiticar, M.J., Faber, E., Schoell, M., 1986. Biogenic methane formation in marine and freshwater environments: CO2 reduction vs. acetate fermentation-isotopic evidence. Geochimica et Cosmochimica Acta 50, 693–709.

75. Widdel, F., 1986. Growth of methanogenic bacteria in pure culture with 2-propanol and other alcohols as hydrogen donors. Applied Environmental Microbiology 51 (5), 1056–1062.

76. Williams, R.T., Crawford, R.L., 1985. Methanogenic bacteria, including an acid-tolerant strain, from peatlands. Applied Environmental Microbiology 50 (6), 1542–1544.

77. Xia, X., 2000. Data Analysis in Molecular Biology and Evolution. Kluwer Academic Publishers, Boston.

78. Zeikus, J.G., Winfrey, M.R., 1976. Temperature limitation of methanogenesis in aquatic sediments. Applied Environmental Microbiology 31 (1), 99–107.

79. Zinder, S.H., 1993. Physiological ecology of methanogens. In: Ferry, J.G. (Ed.), Methanogenesis: Ecology, Physiology, Biochemistry and Genetics. Chapman and Hall, New York, pp. 128–206.

Citations

CHAPTER 1

Jingyu Jiang and Yuanping Cheng, "Effects of Igneous Intrusion on Microporosity and Gas Adsorption Capacity of Coals in the Haizi Mine, China," The Scientific World Journal, vol. 2014, Article ID 976582, 12 pages, 2014. doi:10.1155/2014/976582.

CHAPTER 2

Haipeng Wang, Yushuang Yang, Jianli Yang, Yihang Nie, Jing Jia, and Yudan Wang, "Evaluation of Multiple-Scale 3D Characterization for Coal Physical Structure with DCM Method and Synchrotron X-Ray

CT," The Scientific World Journal, vol. 2015, Article ID 414262, 6 pages, 2015. doi:10.1155/2015/414262.

CHAPTER 3

A. Belhadi, M. Trari, C. Rabia and O. Cherifi, "Methane Steam Reforming on Supported Nickel Based Catalysts. Effect of Oxide ZrO2, La2O3 and Nickel Composition," Open Journal of Physical Chemistry, Vol. 3 No. 2, 2013, pp. 89-96. doi: 10.4236/ojpc.2013.32011.

CHAPTER 4

Rico, C. , Diego, R. , Valcarce, A. and Rico, J. (2014) Biogas Production from Various Typical Organic Wastes Generated in the Region of Cantabria (Spain): Methane Yields and Co-Digestion Tests. Smart Grid and Renewable Energy, 5, 128-136. doi: 10.4236/sgre.2014.56012.

CHAPTER 5

Jack C. Pashin, Marcella R. McIntyre-Redden, Steven D. Mann, David C. Kopaska-Merkel, Matthew Varonka, William Orem, Relationships between water and gas chemistry in mature coalbed methane reservoirs of the Black Warrior Basin, International Journal of Coal Geology, Volume 126, 1 June 2014, Pages 92-105, ISSN 0166-5162, http://dx.doi.org/10.1016/j.coal.2013.10.002.

CHAPTER 6

Yidong Cai, Dameng Liu, Yanbin Yao, Zhentao Li, Zhejun Pan, Partial coal pyrolysis and its implication to enhance coalbed methane recovery, Part I: An experimental investigation, Fuel, Volume 132,

15 September 2014, Pages 12-19, ISSN 0016-2361, http://dx.doi.
org/10.1016/j.fuel.2014.04.084.

CHAPTER 7

Katharine G. Dahm, Katie L. Guerra, Junko Munakata-Marr, Jörg
E. Drewes, Trends in water quality variability for coalbed methane
produced water, Journal of Cleaner Production, Volume 84, 1
December 2014, Pages 840-848, ISSN 0959-6526, http://dx.doi.
org/10.1016/j.jclepro.2014.04.033.

CHAPTER 8

Michael S. Green, Keith C. Flanegan, Patrick C. Gilcrease,
Characterization of a methanogenic consortium enriched from
a coalbed methane well in the Powder River Basin, U.S.A.,
International Journal of Coal Geology, Volume 76, Issues 1–2,
2 October 2008, Pages 34-45, ISSN 0166-5162, http://dx.doi.
org/10.1016/j.coal.2008.05.001.

Index